NUMBER 218

THE ENGLISH EXPERIENCE

ITS RECORD IN EARLY PRINTED BOOKS
PUBLISHED IN FACSIMILE

FRANCIS BACON

THE TWOO BOOKES
OF THE PROFICIENCE
AND ADVANCEMENT OF LEARNING

LONDON 1605

DA CAPO PRESS
THEATRVM ORBIS TERRARVM LTD.
AMSTERDAM 1970 NEW YORK

The publishers acknowledge their gratitude
to the Provost and Fellows of King's College, Cambridge,
for their permission to reproduce
the Library's copy (Shelfmark: Keynes. A.7.9.)
and to the Curators of the Bodleian Library, Oxford,
for their permission to reproduce
the leaves Hhh2 and ¶ 2 from their copy.
(Shelfmark: 4°.B.15.Art.Seld.)

S.T.C. No. 1164
Collation :A — L⁴, M², Aa — Zz⁴, Aaa — Ggg⁴, Hhh², ¶².

Published in 1970 by
Theatrum Orbis Terrarum Ltd.,
O.Z. Voorburgwal 85, Amsterdam
&
Da Capo Press
- a division of Plenum Publishing Corporation -
227 West 17th Street, New York, 10011
Printed in the Netherlands
ISBN 90 221 0218 1

THE
Tvvoo Bookes of
Francis Bacon.

Of the proficience and aduance.
ment of Learning, diuine and
humane.

To the King.

At London,

¶ Printed for *Henrie Tomes*, and
are to be fould at his fhop at Graies Inne
Gate in Holborne. 1605.

THE FIRST BOOKE

of FRANCIS BACON; of the
proficience and aduancement of
Learning diuine, and humane.

To the King.

Here were vnder the Lawe
(excellent King) both dayly
Sacrifices, and free will Of-
ferings ; The one procee-
ding vpon ordinarie obferuance ; The other vppon a
deuout cheerefulneffe : In
like manner there belon-
geth to Kings from their Seruants, both Tribute of
dutie, and prefents of affection : In the former of
thefe, I hope I fhal not liue to be wanting, according
to my moft humble dutie, and the good pleafure of
your Maiefties employments: for the later, I thought
it more refpectiue to make choyce of fome oblation, which might rather referre to the proprietie
and excellencie of your indiuiduall perfon, than to
the bufineffe of your Crowne and State.

Wherefore reprefenting your Maieftie many
times vnto my mind, and beholding you not with

the

the inquiſitiue eye of preſumption, to diſcouer that
which the Scripture telleth me is inſcrutable; but
with the obſeruant eye of dutie and admiration:
leauing aſide the other parts of your vertue and fortune, I haue been touched, yea and poſſeſſed with
an extreame woonder at thoſe your vertues and faculties, which the Philoſophers call intellectuall:
The largeneſſe of your capacitie, the faithfulneſſe
of your memorie, the ſwiftneſſe of your apprehen-
ſion, the penetration of your Iudgement, and the
facilitie and order of your elocution; and I haue
often thought, that of all the perſons liuing, that I
haue knowne, your Maieſtie were the beſt inſtance
to make a man of *Platoes* opinion, that all knowledge is but remembrance, and that the minde of
man by nature knoweth all things, and hath but her
owne natiue and originall motions (which by the
ſtrangeneſſe and darkeneſſe of this Tabernacle of
the bodie are ſequeſtred) againe reuiued and reſtored: ſuch a light of Nature I haue obſerued in your
Maieſtie, and ſuch a readineſſe to take flame, and
blaze from the leaſt occaſion preſented, or the leaſt
ſparke of anothers knowledge deliuered. And as
the Scripture ſayth of the wiſeſt King: *That his heart
was as the ſands of the Sea,* which though it be one of
the largeſt bodies, yet it côſiſteth of the ſmalleſt & fi-
neſt portions: So hath God giuen your Maieſtie a
côpoſition of vnderſtanding admirable, being able
to compaſſe & comprehend the greateſt matters, &
neuertheleſſe to touch and apprehend the leaſt;
whereas

wheras it should seeme an impossibility in Nature, for the same Instrument to make it selfe fit for great and small workes. And for your gift of speech, I call to minde what *Cornelius Tacitus* sayth of *Augustus Cæsar* : *Augusto profluens & quæ principem deceret, eloquentia fuit* : For if we note it well, speech that is vttered with labour and difficultie, or speech that sauoreth of the affectation of art and precepts, or speech that is framed after the imitation of some patterne of eloquence, though neuer so excellent: All this hath somewhat seruile, and holding of the subiect. But your Maiesties manner of speech is indeed Prince-like, flowing as from a fountaine, and yet streaming & branching it selfe into Natures order, full of facilitie, & felicitie, imitating none & ininimitable by any. And as in your ciuile Estate there appeareth to be an emulation & contentiõ of your Maiesties vertue with your fortune, a vertuous disposition with a fortunate regiment, a vertuous expectation (when time was) of your greater fortune, with a prosperous possession thereof in the due time; a vertuous obseruation of the lawes of marriage, with most blessed and happie fruite of marriage; a vertuous and most christian desire of peace, with a fortunate inclination in your neighbour Princes thereunto; So likewise in these intellectuall matters, there seemeth to be no lesse contention betweene the excellencie of your Maiesties gifts of Nature, and the vniuersalitie and profection of your learning For I am well assured, that this which

I shall say is no amplification at all, but a positiue and meafured truth: which is, that there hath not beene fince Chrifts time any King or temporall Monarch which hath ben fo learned in all literature & erudition, diuine & humane. For let a man ferioufly & diligently reuolue and perufe the fuccefsion of the Emperours of Rome, of which *Cæfar* the Dictator, who liued fome yeeres before Chrift, and *Marcus Antoninus* were the beft learned: and fo defcend to the Emperours of *Grecia*, or of the Weft, and then to the lines of *Fraunce, Spaine, England, Scotland* and the reft, and he fhall finde this iudgement is truly made. For it feemeth much in a King, if by the compendious extractions of other mens wits and labours, he can take hold of any fuperficiall Ornaments and fhewes of learning, or if he countenance and preferre learning and learned men: But to drinke indeed of the true Fountaines of learning, nay, to haue fuch a fountaine of learning in himfelfe, in a King, and in a King borne, is almoft a Miracle. And the more, becaufe there is met in your Maiefty a rare Coniunction, afwell of diuine and facred literature, as of prophane and humane; So as your Maieftie ftandeth inuefted of that triplicitie, which in great veneration, was afcribed to the ancient *Hermes*; the power and fortune of a King; the knowledge and illumination of a Prieft; and the learning and vniuerfalitie of a Philofopher. This propriety inherent and indiuiduall attribute in your Maieftie deferueth to be expreffed, not onely in the fame and

admiration

admiration of the prefent time, nor in the Hiflorie or tradition of the ages fucceeding; but alfo in fome folide worke, fixed memoriall, and immortall monument, bearing a Character or fignature, both of the power of a king, and the difference and perfection of fuch a king.

Therefore I did conclude with my felfe, that I could not make vnto your Maiefty a better oblation, then of fome treatife tending to that end, whereof the fumme will confift of thefe two partes: The former concerning the excellencie of learning and knowledge, and the excellencie of the merit and true glory, in the Augmentation and Propagation thereof: The latter, what the particuler actes and workes are, which haue been imbraced and vndertaken for the aduancement of learning : And againe what defects and vndervalewes I finde in fuch particuler actes: to the end, that though I cannot pofitiuely or affirmatiuelie aduife your Maieftie, or propound vnto you framed particulers; yet I may excite your princely Cogitations to vifit the excellent treafure of your owne mind, and thence to extract particulers for this purpofe, agreeable to your magnanimitie and wifedome.

IN the entrance to the former of thefe ; to cleere the way, & as it were to make filence, to haue the true teftimonies concerning the dignitie of Learning to be better heard, without the interruption of tacite obiections ; I thinke good to deliuer it from

the

the difcredites and difgraces which it hath receiued; all from ignorance; but ignorance feuerally
difguifed, appearing fometimes in the zeale and iea-
loufie of Diuines; fometimes in the feueritie and
arrogancie of Politiques, and fometimes in the errors and imperfe&ions of learned men themfelues.

I heare the former fort fay, that knowledge is of
thofe things which are to be accepted of with great
limitation and caution, that th'afpiring to ouermuch knowledge, was the originall temptation and
finne, whereupon enfued the fal of Man that knowledge hath in it fomewhat of the Serpent, and therefore where it entreth into a man, it makes him fwel.
Scientia inflat. That *Salomon* giues a Cenfure,
*That there is no end of making Bookes, and that much
reading is wearineffe of the flefh.* And againe in another
place, *That in fpatious knowledge, there is much contri-
ftation, and that he that encreafeth knowledge, encreafeth
anxietie:* that Saint *Paul* giues a Caueat, *that we be
not fpoyled through vaine Philofophie:* that experience demonftrates, how learned men, haue beene
Arch-heretiques, how learned times haue been enclined to Atheifme, and how the contemplation of
fecond Caufes doth derogate frō our dependance
vpon God, who is the firft caufe.

To difcouer then the ignorance & error of this o-
pinion, and the mifunderftanding in the grounds
thereof, it may well appeare thefe men doe not obferue or cōfider, that it was not the pure knowledg
of nature and vniuerfality, a knowledge by the light
whereof

whereof man did giue names vnto other creatures
in Paradife, as they were brought before him, according vnto their proprieties, which gaue the occafion to the fall ; but it was the proude knowledge
of good and euill, with an intent in man to giue law
vnto himfelfe, and to depend no more vpon Gods
commaundements, which was the fourme of the
temptation; neither is it any quantitie of knowledge
how great foeuer that can make the minde of man
to fwell; for nothing can fill, much leffe extend the
foule of man, but God, and the contemplation of
God; and therfore *Salomon* fpeaking of the two principall fences of Inquifition, the Eye, and the Eare, af-
firmeth that the Eye is neuer fatisfied with feeing,
nor the Eare with hearing ; and if there be no fulneffe, then is the Continent greater, than the Content ; fo of knowledge it feife, and the minde of
man, whereto the fences are but Reporters, he
defineth likewife in thefe wordes, placed after
that Kalender or Ephemerides, which he maketh
of the diuerfities of times and feafons for all actions and purpofes ; and concludeth thus : *God
hath made all thinges beautifull or decent in the true
returne of their feafons . Alfo hee hath placed the
world in Mans heart , yet cannot Man finde out the
worke which God worketh from the beginning to the
end*: Declaring not obfcurely, that God hath framed the minde of man as a mirrour, or glaffe, capable of the Image of the vniuerfall world, and
ioyfull to receiue the imprefsion thereof, as the

Eye

Eye ioyeth to receiue light, and not onely delighted in beholding the varietie of thinges and
vicissitude of times, but rayfed alfo to finde out
and difcerne the ordinances and decrees which
throughout all thofe Changes are infallibly obferued. And although hee doth infinuate that
the fupreame or fummarie law of Nature, wh ch
he calleth, *The worke which God worketh from the
beginning to the end, is not pofsible to be found out
by Man;* yet that doth not derogate from the
capacitie of the minde ; but may bee referred
to the impediments as of fhortneffe of life, ill
coniunction of labours, ill tradition of knowledge ouer from hand to hand, and many other
Inconueniences, whereunto the condition of
Man is fubiect. For that nothing parcell of the
world, is denied to Mans enquirie and inuention : hee doth in another place rule ouer; when
hee fayth, *The Spirite of Man is as the Lampe of
God, wherewith hee fearcheth the inwaranneffe of
all fecrets.* If then fuch be the capacitie and receit of the mind of Man, it is manifeft, that there
is no daunger at all in the proportion or quantitie of knowledge howe large foeuer ; leaft it
fhould make it fwell or outcompaffe it felfe ; no,
but it is meerely the qualitie of knowledge,
which be it in quantitie more or leffe, if it bee
taken without the true correctiue thereof, hath
in it fome Nature of venome or malignitie, and
fome effects of that venome which is ventofitie

or

or fwelling. This correctiue fpice, the mixture whereof maketh knowledge fo foueraigne, is Charitie, which the Apoftle imediately addeth to the former claufe for fo he fayth, *Knowledge bloweth vp, but Charitie buildeth vp* ; not vnlike vnto that which hee deliuereth in another place : *If I fpake (fayth hee) with the tongues of men and Angels, and had not Charitie, it were but as a tinckling Cymball*; not but that it is an excellent thinge to fpeake with the tongues of Men and Angels, but becaufe if it bee feuered from Charitie, and not referred to the good of Men and Mankind, it hath rather a founding and vnworthie glorie, than a meriting and fubftantiall vertue. And as for that Cenfure of *Salomon* concerning the exceffe of writing and reading Bookes, and the anxietie of fpirit which redoundeth from knowlegde, and that admonition of Saint *Paule, That wee bee not feduced by vayne Philofophie*; let thofe places bee rightly vnderftoode, and they doe indeede excellently fette foorth the true bounds and limitations, whereby humane knowledge is confined and circumfcribed : and yet without any fuch contracting or coarctation, but that it may comprehend all the vniuerfall nature of thinges : for thefe limitations are three : The firft, *That wee doe not fo place our felicitie in knowledge, as wee forget our mortalitie.* The fecond, *that we make application of our knowledge to giue our felues repofe and contentment, and not diftaft or repining.*

B 2   The

The third: that we doe not prefume by the contemplation of Nature, to attaine to the mifteries of
God ; for as touching the firft of thefe, *Salomon* doth
excellently expound himfelfe in another place of
the fame Booke, where hee fayth ; *I fawe well
that knowledge recedeth as farre from ignorance, as
light doth from darkeneffe, and that the wife mans
eyes keepe watch in his head whereas the foole roun-
deth about in darkeneffe : But withall I learned that
the fame mortalitie inuolueth them both.* And for
the fecond, certaine it is, there is no vexation or anxietie of minde, which refulteth from knowledge o-
therwife than meerely by accident ; for all knowledge and wonder (which is the feede of knowledge) is an imprefsion of pleafure in it felfe; but
when men fall to framing conclufions out of their
knowledge, applying it to their particuler, and
miniftring to themfelues thereby weake feares,
or vaft defires, there groweth that carefulnefle
and trouble of minde, which is fpoken of: for then
knowledge is no more *Lumen ficcum*, whereof
Heraclitus the profound fayd, *Lumen ficcum opti-
ma anima*, but it becommeth *Lumen madidum*, or *ma-
ceratum*, being fteeped and infufed in the humors of
the affections. And as for the third point, it deferueth to be a little ftood vpon, and not to be lightly
paffed ouer : for if any man fhall thinke by view and
enquiry into thefe fenfible and material things to attaine that light, whereby he may reueale vnto himfelfe the nature or will of God : then indeed is he

fpoyled

spoyled by vaine Philofophie: for the contemplation of Gods Creatures and works produceth (hauing regard to the works and creatures themfelues)
knowledge, but hauing regard to God, no perfect
knowledg, but wonder, which is broke knowledge:
And therefore it was moft aptly fayd by one of *Pla-*
toes Schoole, *That the fence of man caryeth a refem-*
blance with the Sunne, which (as we fee) openeth and re-
uealeth all the terreftriall Globe; but then againe it obfcu-
reth and concealeth the ftars & celeftiall Globe : So doth
the Sence difcouer naturall thinges, but it darkeneth and
fhutteth vp Diuine. And hence it is true that it
hath proceeded that diuers great learned men
haue beene hereticall, whileft they haue fought
to flye vp to the fecrets of the Deitie by the
waxen winges of the Sences : And as for the
conceite that too much knowledge fhould encline a man to Atheifme, and that the ignorance
of fecond caufes fhould make a more deuoute
dependance vppon God, which is the firft caufe;
Firft, it is good to afke the queftion which *Iob*
asked of his friends ; *Will you lye for God, as one*
man will doe for another, to gratifie him ? for certaine
it is, that God worketh nothing in Nature, but by
fecond caufes, and if they would haue it otherwife
beleeued, it is meere impofture, as it were in fauour
towardes God ; and nothing elfe, but to offer to the
Author of truth, the vncleane facrifice of a lye.
But further, it is an affured truth, and a conclufion of experience, that a little or fuperficiall

Of the aduancement of learning,

knowledge of Philofophie may encline the minde of Man to Atheifme, but a further proceeding therem doth bring the mind backe againe to Religion: for in the entrance of Philofophie, when the fecond Caufes, which are next vnto the fences, do offer themfelues to the minde of Man, if it dwell and ftay there, it may induce fome obliuion of the higheft caufe; but when a man paffeth on further, and feeth the dependance of caufes, and the workes of prouidence; then according to the allegorie of the Poets, he will eafily beleeue that the higheft Linke of Natures chaine muft needes be tyed to the foote of *Iupiters* chaire. To conclude therefore, let no man vppon a weake conceite of fobrietie, or an ill applyed moderation thinke or maintaine, that a man can fearch too farre, or bee too well ftudied in the Booke of Gods word, or in the Booke of Gods workes; Diuinitie or Philofophie; but rather let men endeauour an endlefle progreffe or proficience in both: only let men beware that they apply both to Charitie, and not to fwelling; to vfe, and not to oftentation; and againe, that they doe not vnwifely mingle or confound thefe learnings together.

And as for the difgraces which learning receiueth from Politiques, they bee of this nature; that learning doth foften mens mindes, and makes them more vnapt for the honour and exercife of Armes; that it doth marre and peruert mens difpofitions for

matter

matter of gouernement and policie ; in making
them too curious and irresolute by varietie of reading; or too peremptorie or positiue by stricktnesse
of rules and axiomes; or too immoderate and o-
uerweening by reason of the greatnesse of examples; or too incompatible and differing from the
times, by reason of the dissimilitude of examples;
or at least, that it doth diuert mens trauai es from action and businesse, and bringeth them to a loue of
leasure and priuatenesse ; and that it doth bring into States a relaxation of discipline, whilst euerie
man is more readie to argue, than to obey
and execute. Out of this conceit, *Cato* surnamed
the Censor, one of the wisest men indeed that euer
liued, when *Carneades* the Philosopher came in
Embassage to Rome, and that the young men of
Rome began to flocke about him, being allured
with the sweetnesse and Maiestie of his eloquence
and learning, gaue counsell in open Senate, that
they should giue him his dispatch with all speede,
least hee should infect and inchaunt the mindes
and affections of the youth, and at vnawares bring
in an alteration of the manners and Customes
of the State. Out of the same conceite or humor
did *Virgill*, turning his penne to the aduantage
of his Countrey, and the disaduantage of his owne
profefsion, make a kind of separation betweene policie and gouernement, and betweene Arts and Sciences, in the verses so much renowned, attributing and challenging the one to the Romanes, and

B 4   leauing

leauing & yeelding the other to the Grecians, *Tu regere imperio populos Romane memēto, Hæ tibi erūt artes,* &c. so likewise we see that *Anytus* the accuser of *Socrates* layd it as an Article of charge & accusation against him, that he did with the varietie and power of his discourses and disputations withdraw young men from due reuerence to the Lawes and Customes of their Countrey: and that he did professe a dangerous aud pernitious Science, which was to make the worse matter seeme the better, and to suppresse truth by force of eloquence and speech.

But these and the like imputations haue rather a countenance of grauitie, than any ground of Iustice: for experience doth warrant, that both in persons and in times, there hath beene a meeting, and concurrence in learning and Armes, flourishing and excelling in the same men, and the same ages. For as for men, there cannot be a better nor the like instance, as of that payre *Alexander* the Great, and *Iulius Cæsar* the Dictator, whereof the one was *Aristotles* Scholler in Philosophie, and the other was *Ciceroes* Riuall in eloquence; or if any man had rather call for Schollers, that were great Generals, then Generals that were great Schollers; let him take *Epaminondas* the Thebane, or *Xenophon* the Athenian, whereof the one was the first that abated the power of *Sparta*; and the other was the first that made way to the ouerthrow of the Monarchie of *Persia*: And this concurrence is yet more visible in times than in persons, by how much an age is greater object ieɛt than a Man. For both in *Ægypt*, *Afsyria*, *Per-fia*, *Grecia*, and *Rome* the same times that are most re-nowned for Armes, are likewise most admired for learning ; so that the greatest Authors and Philoso-phers, and the greatest Captaines and Gouernours haue liued in the same ages: neither can it other-wise be; for as in Man, the ripenesse of strength of the bodie and minde commeth much about an age, saue that the strength of the bodie commeth somewhat the more early ; So in States, Armes and Learning, whereof the one correspondeth to the bodie, the other to the soule of Man, haue a concur-rence or nere sequence in times.

And for matter of policie and gouernement, that Learning should rather hurt, than inable thereunto, is a thing verie improbable : we see it is accounted an errour, to commit a naturall bodie to Empe-rique Phisitions, which commonly haue a fewe pleasing receits, whereupon they are confident and aduenturous, but know neither the causes of dis-eases, nor the complexions of Patients, nor perill of accidents, nor the true methode of Cures; We see it is a like error to rely vpon Aduocates or Law-yers, which are onely men of practise, and not grounded in their Bookes, who are many times ea-sily surprised, when matter falleth out besides their experience, to the preiudice of the causes they han-dle: so by like reason it cannot be but a matter of doubtfull consequence, if States bee managed by

C   Empe-

Emperique Statefmen, not well mingled with men grounded in Learning. But contrary wife, it is almoft without inftance contradictorie, that euer any gouernement was difaftrous that was in the hands of learned Gouernors. For howfoeuer it hath beene ordinarie with politique men to extenuate and difable learned men by the names of *Pedantes* : yet in the Records of time it appeareth in many particulers, that the Gouernements of Princes in minority(notwithftanding the infinite difaduantage of that kinde of State) haue neuerthelefle excelled the gouernement of Princes of mature age, euen for hat reafon, which they feek to traduce, which is, that by that occafion the ftate hath been in the hands of *Pedantes* : for fo was the State of Rome for the firft fiue yeeres, which are fo much magnified, during the minoritie of *Nero*, in the handes of *Seneca* a *Peaanti:* So it was againe for ten yeres fpace or more during the minoritie of *Gordianus* the younger, with great applaufe and contentation in the hands of *Mifitheus* a *Pedanti* : fo was it before that, in the minoritie of *Alexander Seuerus* in like happinefle, in hands not much vnlike, by reafon of the rule of the women, who were ayded by the Teachers and Preceptors. Nay, let a man locke into the gouernement of the Bi hops of Rome, as by na ne, into the gouernement of *Pius Quintus*, and *Sextus Quintus* in our times, who were both at their entrance efteemed, but as Pedanticall Friers and he fhall find that fuch Popes doe greater thinges, and proceed vpon truer

principles

principles of Estate, than those which haue ascended to the Papacie from an education & breeding
in affaires of Estate, and Courts of Princes; for
although men bred in Learning,are perhaps to seeke
in points of conuenience, and accommodating for
the present which the Italians call *Ragioni di stato*,
whereof the same *Pius Quintus* could not heare
spoken with patience, tearming them Inuentions a-
gainst Religion and the morall vertues; yet on the
other side to recompence that,they are perfite in
those same plaine grounds of Religion, Iustice, Honour,and Morall vertue; which if they be well and
watchfully pursued,there will bee seldome vse of
those other, no more than of Phisicke in a sound or
well dieted bodie; neither can the experience of
one mans life, furnish examples and presidents for
the euents of one mans life. For as it happeneth
sometimes, that the Graund child, or other descen-
dent,resembleth the Ancestor more than the Sonne:
so many times occurrences of present times may
sort better with ancient examples, than with those
of the later or immediate times;and lastly,the wit of
one man. can no more counteruaile learning, than
one mans meanes can hold way with a common
purse.

And as for those particular seducements or dis-
positions of the minde for policie and gouernement, which learning is pretended to insinuate; if it
be graunted that any such thing be, it must be remembred withall, that learning ministreth in euery

of them greater ſtrength of medicine or remedie, than it offereth cauſe of indiſpoſition or infirmitie; For if by a ſecret operation, it make men perplexed and irreſolute, on the other ſide by plaine precept, it teacheth them when, and vpon what ground to reſolue: yea, and how to carrie thinges in ſuſpence without preiudice, till they reſolue: If it make men poſitiue and reguler, it teacheth them what thinges are in their nature demonſtratiue, & what are conieɛturall; and aſwell the vſe of diſtinɛtions, and exceptions, as the latitude of principles and rules. If it miſlead by diſproportion, or diſsimilitude of Examples, it teacheth men the force of Circumſtances, the errours of compariſons, and all the cautions of application: ſo that in all theſe it doth reɛtifie more effeɛtually,than it can peruert.And theſe medicines it conueyeth into mens minds much more forcibly by the quickneſſe and penetration of Examples: for let a man looke into the errours of *Clement* the ſeuenth, ſo liuely deſcribed by *Guicciardine,* who ſerued vnder him, or into the errours of *Cicero* painted out by his owne penſill in his Epiſtles to *Atticus,* and he will flye apace from being irreſolute. Let him looke into the errors of *Phocion,* and he will beware how he be obſtinate or inflexible Let him but read the Fable of *Ixion,* and it will hold him from being vaporous or imaginatiue; let him look into the errors of *Cato* the ſecond, and he will neuer be one of the *Antipodes,* to tread oppoſite to the preſent world.

And for the conceite that Learning ſhould diſpoſe

poſe men to leaſure and priuateneſſe, and make
men ſlouthfull: it were a ſtrange thing if that which
accuſtometh the minde to a perpetuall motion and
agitation, ſhould induce ſlouthfulneſſe, whereas
contrariwiſe it may bee truely affirmed, that no
kinde of men loue buſineſſe for it ſelfe, but thoſe
that are learned; for other perſons loue it for profite; as an hireling that loues the worke for the wa-
ges; or for honour; as becauſe it beareth them vp
in the eyes of men, and refreſheth their reputation,
which otherwiſe would weare; or becauſe it putteth them in mind of their fortune, and giueth them
occaſion to pleaſure and diſpleaſure; or becauſe it
exerciſeth ſome faculty, wherein they take pride,
and ſo entertaineth them in good humor, and plea-
ſing conceits toward themſelues; or becauſe it ad-
uanceth any other their ends. So that as it is ſayd of
vntrue valors, that ſome mens valors are in the eyes
of them that look on; So ſuch mens induſtries are in
the eyes of others, or at leaſt in regard of their owne
deſignements; onely learned men loue buſineſſe, as
an action according to nature, as agreable to health
of minde, as exerciſe is to health of bodie, taking
pleaſure in the action it ſelfe, & not in the purchaſe:
So that of all men, they are the moſt indefatigable,
if it be towards any buſineſſe, which can hold or de-
taine their minde.

And if any man be laborious in reading and ſtudy,
and yet idle in buſines & action, it groweth frō ſome
weakenes of body, or ſoftnes of ſpirit; ſuch as *Seneca*

ſpeaketh

speaketh of: *Quidam tam sunt vmbratiles, vt putent in turbido esse, quicquid in luce est* ; and not of learning ; wel may it be that such a point of a mans nature may make him giue himselfe to learning, but it is not learning that breedeth any such point in his Nature.

And that learning should take vp too much time or leasure, I answere, the most actiue or busie man that hath been or can bee, hath (no question) many vacant times of leasure, while he expecteth the tides and returnes of businesse (except he be either tedious, and of no dispatch, or lightly and vnworthily ambitious, to meddle in thinges that may be better done by others) and then the question is, but how those spaces and times of leasure shall be filled and spent: whether in pleasures, or in studies ; as was wel answered by *Demosthenes* to his aduersarie *Æschynes*, that was a man giuen to pleasure, and told him, *That his Orations did smell of the Lampe: Indeede* (sayd *Demosthenes*) *there is a great difference betweene the thinges that you and I doe by Lampe-light :* so as no man neede doubt, that learning will expulse businesse, but rather it will keepe and defend the possession of the mind against idlenesse and pleasure, which otherwise, at vnawares, may enter to the preiudice of both.

Againe, for that other conceit, that learning should vndermine the reuerence of Lawes and gouernement, it is assuredly a meere deprauation and calumny without all shadowe of truth : for to say that a blind custome of obedience should be a surer obligation

ligation, than dutie taught and vnderstood; it is to affirme that a blind man may tread surer by a guide, than a seeing man can by a light: and it is without all controuersie, that learning doth make the minds of men gentle, generous, maniable, and pliant to gouernment, whereas Ignorance makes them churlish thwart, and mutinous; and the euidence of time doth cleare this assertion, considering that the most barbarous, rude, and vnlearned times haue beene most subiect to tumults, seditions, and changes.

And as to the iudgement of *Cato* the Censor, he was well punished for his blasphemie against learning in the same kinde wherein hee offended; for when he was past threescore yeeres old, he was taken with an extreame desire to goe to Schoole againe, and to learne the Greeke tongue, to the end to peruse the Greeke Authors; which doth well demonstrate, that his former censure of the Grecian Learning, was rather an affected grauitie, than according to the inward sence of his owne opinion. And as for *Virgils* verses, though it pleased him to braue the world in taking to the Romanes, the Art of Empire, and leauing to others the arts of subiects: yet so much is manifest, that the Romanes neuer ascended to that height of Empire, till the time they had ascended to the height of other Arts: For in the time of the two first *Cæsars*, which had the art of gouernement in greatest perfection, there liued the best Poet *Virgilius Maro*, the best Historiographer *Titus Liuius*, the best Antiquarie *Marcus Varro*, and
C 4　　　　the

the beſt or ſecond Orator *Marcus Cicero,* that to the memorie of man are knowne. As for the accuſation of *Socrates,* the time muſt be remembred, when it was proſecuted ; which was vnder the thirtie Tyrants, the moſt baſe, bloudy, and enuious perſons that haue gouerned ; which reuolution of State was no ſooner ouer, but *Socrates,* whom they had made a perſon criminall, was made a perſon heroycall, and his memorie accumulate with honors diuine and humane ; and thoſe diſcourſes of his which were then tearmed corrupting of manners, were after acknowledged for ſoueraigne Medicines of the minde and manners, and ſo haue beene receiued euer ſince till this day. Let this therefore ſerue for anſwere to Politiques, which in their humorous ſeueritie, or in their fayned grauitie haue preſumed to throwe imputations vpon learning, which redargution neuertheleſſe (ſaue that wee know not whether our labours may extend to other ages) were not needefull for the preſent, in regard of the loue and reuerence towards Learning, which the example and countenance of twoo ſo learned Princes Queene *Elizabeth,* and your Maieſtie ; being as *Caſtor* and *Pollux, Lucida Sydera,* Starres of excellent light, and moſt benigne influence, hath wrought in all men of place and authoritie in our Nation.

Now therefore, we come to that third ſort of diſcredite, or diminution of credite, that groweth vnto learning from learned men themſelues, which commonly cleaueth faſteſt ; It is either from their

fortune,

Fortune, or frō their Manners, or from the nature of their Studies: for the firſt, it is not in their power; and the ſecond is accidentall; the third only is proper to be hādled but becauſe we are not in hand with true meaſure, but with popular eſtimation & conceit, it is not amiſſe to ſpeak ſomwhat of the two former. Tho derogations therfore, which grow to learning from the fortune or condition of learned men, are either in reſpeċt of ſcarſity of meanes, or in reſpeċt of priuateneſſe of life, and meaneſſe of employments.

Concerning want, and that it is the caſe of learned men, vſually to beginne with little, and not to growe rich ſo faſt as other men, by reaſon they conuert not their labors chiefely to luker, and encreaſe; It were good to leaue the common place in commendation of pouertie to ſome Frier to handle, to whom much was attributed by *Macciauell* in this point, when he ſayd, *That the Kingdome of the Clear-gie had beene long before at an end, if he reputation and reuerence towards the pouertie of Friers had not borne out the ſcandall of the ſuperfluities and exceſſes of Bi-ſhops and Prelates.* So a man might ſay, that the felicitie and delicacie of Princes and great Perſons, had long ſince turned to Rudenes and Barbariſme, if the pouertie of Learning had not kept vp Ciuilitie and Honor of life; But without any ſuch aduantages, it is worthy the obſeruation, what a reuerent and honoured thing pouertie of fortune was, for ſome ages in the Romane State, which neuertheleſſe was a State without paradoxes. For we ſee what *Titus Li-*

uius sayth in his introduction. *Cæterum aut me amor negotij suscepti fallit, aut nulla vnquam respublica, nec maior, nec sanctior, nec bonis exemplis ditior fuit ; nec in quam tam seræ auaritia luxuriáque immigrauerint, nec vbi tantus ac tam diu paupertati ac parsimoniæ honos fuerit.* We see likewise after that the State of Rome was not it selfe, but did degenerate ; how that person that tooke vpon him to be Counsellor to *Iulius Cæsar,* after his victorie, where to begin his restauration of the State, maketh it of all points the most summarie to take away the estimation of wealth. *Verum hæc & omnia mala pariter cum honore pecuniæ desinent ; Si neque Magistratus, neque alia vulgo cupienda venalia erunt.* To conclude this point, as it was truely sayd, that *Rubor est virtutis color,* though sometime it come from vice: So it may be fitly sayd, that *Paupertas est virtutis fortuna.* Though sometimes it may proceede from misgouernement and accident. Sureq *Salomon* hath pronounced it both in censure, *Qui festinat ad diuitias non erit insons;* and in precept : *Buy the truth, and sell it not: and so of wisedome and knowledge;* iudging that meanes were to be spent vpon learning, and not learning to be applyed to meanes: And as for the priuatenesse or obscurenesse (as it may be in vulgar estimation accounted) of life of contemplatiue men ; it is a Theame so common, to extoll a priuate life, not taxed with sensualitie and sloth in comparison, and to the disaduantage of a ciuile life, for safety, libertie, pleasure and dignitie, or at least freedome from indignitie,

tie, as no man handleth it, but handleth it well: such a confonancie it hath to mens conceits in the exprefsing, and to mens confents in the allowing : this onely I will adde ; that learned men forgotten in States, and not liuing in the eyes of men, are like the Images of *Caßius* and *Brutus* in the funerall of *Iunia*; of which not being reprefented, as many others were *Tacitus* fayth, *Eo ipfo prefulgebant, quod non vifebantur.*

And for meaneffe of employment, that which is moft traduced to contempt, is that the gouernment of youth is commonly allotted to them, which age, becaufe it is the age of leaft authoritie, it is transferred to the difefteeming of thofe employments wherin youth is conuerfant, and which are conuerfant about youth. But how vniuft this traducement is, (if you will reduce thinges from popularitie of opinion to meafure of reafon) may appeare in that we fee men are more curious what they put into a new Veffell, than into a Veffell feafoned ; and what mould they lay about a young plant, than about a Plant corroborate; fo as the weakeft Termes and Times of all things vfe to haue the beft applications and helpes. And will you hearken to the Hebrew *Rabynes? Your young men fhall fee Vifions, and your old men fhal dreame dreames,* fay they youth is the worthier age, for that Vifions are neerer apparitions of God, than dreames? And let it bee noted, that how foeuer the Conditions of life of *Pedantes* hath been fcorned vpon Theaters, as the Ape of Tyrannie;

and

and that the modern loofenes or negligence hath taken no due regard to the choife of ſchoolemaſters, & Tutors; yet the ancient wiſdome of the beſt times did alwaies make a iuſt complaint; that States were too buſie with their Lawes, and too negligent in point of education : which excellent part of ancient diſcipline hath been in ſome ſort reuiued of late times, by the Colledges of the Ieſuites: of whom, although in regard of their ſuperſtition I may ſay, *Quo meliores, eo deteriores,* yet in regard of this, and ſome other points concerning humane learning, and Morall matters, I may ſay as *Ageſilaus* ſayd to his enemie *Farnabaſus, Talis quum ſis, vtinam noſter eſſes.* And thus much touching the diſcredits drawn from the fortunes of learned men.

As touching the Manners of learned men, it is a thing perſonall and indiuiduall, and no doubt there be amongſt them, as in other profeſsions, of all temperatures; but yet ſo as it is not without truth, which is ſayd, that *Abeunt ſtudia in mores,* Studies haue an influence and operation, vpon the manners of thoſe that are conuerſant in them.

But vpon an attentiue, and indifferent reuiew; I for my part, cannot find any diſgrace to learning, can proceed frō the manners of learned men; not inherent to them as they are learned; except it be a fault, (which was the ſuppoſed fault of *Demoſthenes, Cicero, Cato* the ſecond, *Seneca,* and many moe) that becauſe the times they read of, are commonly better than the times they liue in and the duties taught,

better

better than the duties practised: They contend somtimes too farre, to bring thinges to perfection; and
to reduce the corruption of manners, to honestie of
precepts, or examples of too great height; And yet
hereof they haue Caueats ynough in their owne
walkes: For *Solon,* when he was asked whether he
had giuen his Citizens the best laws, answered wisely, *Yea of such, as they would receiue*: and *Plato* finding
that his owne heart, could not agree with the corrupt manners of his Country, refused to beare place
or office, saying: *That a mans Countrey was to be vsed
as his Parents were, that is, with humble perswasions, and
not with contestations.* And *Cæsars* Counsellor put in
the same Caueat, *Non ad vetera instituta reuocans quæ
iampridem corruptis moribus ludibrio sunt*; and *Cicero*
noteth this error directly in *Cato* the second, when
he writes to his friend *Atticus; Cato optimè sentit,
sed nocet interdum reipublicæ; loquitur enim tanquam in
repub: Platonis, non tanquam in fæce Romuli*; and the
same *Cicero* doth excuse and expound the Philosophers for going too far, and being too exact in their
prescripts, when he saith; *Isti ipsi præceptores virtutis & Magistri, videntur fines officiorum paulo longius
quam natura vellet protulisse, vt cū ad vltimū animo con-
tendissemus, ibi tamcn, vbi oportet, consisteremus*: and yet
himself might haue said: *Monitis sū minor ipse meis*, for
it was his own fault, thogh not in so extream a degre.

Another fault likewise much of this kind, hath
beene incident to learned men; which is that they
haue esteemed the preseruation, good, and honor of

their

their Countreys or Maifters before their owne fortunes or fafeties. For fo fayth *Demofthenes* vnto the Athenians; *If it pleafe you to note it, my counfels vnto you, are not fuch, whereby I fhould grow great amongft you, and you become little amongft the Grecians: But they be of that nature as they are fometimes not good for me to giue, but are alwaies good for you to follow.* And fo *Seneca* after he had confecrated that *Quinquennium Neronis* to the eternall glorie of learned Gouernors, held on his honeft and loyall courfe of good and free Counfell, after his Maifter grew extreamely corrupt in his gouernment; neither can this point otherwife be: for learning endueth mens mindes with a true fence of the frailtie of their perfons, the cafualtie of their fortunes, and the dignitie of their foule and vocation; fo that it is impofsible for them to efteeme that any greatneffe of their owne fortune can bee, a true or worthy end of their being and ordainment; and therefore are defirous to giue their account to God, and fo likewife to their Maifters vnder God (as Kinges and the States that they ferue) in thefe words; *Ecce tibi lucrifeci,* and not *Ecce mihi lucrifeci*: whereas the corrupter fort of meere Politiques, that haue not their thoughts eftablifhed by learning in the loue and apprehenfion of dutie, nor neuer looke abroad into vniuerfalitie; doe referre all thinges to themfelues, and thruft themfelues into the Center of the world, as if all lynes fhould meet in them and their fortunes; neuer caring in all tempefts what becoms of the fhippe of Eftates,

Eſtates, ſo they may ſaue themſelues in the Cocke-
boat of their owne fortune, whereas men that feele
the weight of dutie, and know the limits of ſelfe-
loue, vſe to make good their places & duties, though
with perill. And if they ſtand in ſeditious and vio-
lent alterations ; it is rather the reuerence which
many times both aduerſe parts doe giue to honeſtie,
than any verſatile aduantage of their owne carriage.
But for this point of tender fence, and faſt obligation of dutie, which learning doth endue the minde
withall, howſoeuer fortune may taxe it, and many
in the depth of their corrupt principles may deſpiſe
it, yet it will receiue an open allowance, and therefore needes the leſſe diſproofe or excuſation.

Another fault incident commonly to learned
men, which may be more probably defended, than
truely denyed, is ; that they fayle ſometimes in applying themſelus to particular perſons, which want
of exact application ariſeth from two cauſes : The
one, becauſe the largeneſſe of their minde can hardly confine it ſelfe to dwell in the exquiſite obſeruation or examination of the nature and cuſtomes of
one perſon : for it is a ſpeech for a Louer, & not for a
wiſe man : *Satis magnum alter alteri Theatrum ſumus*:
Neuertheleſſe I ſhall yeeld, that he that cannot contract the ſight of his minde, aſwell as diſperſe and
dilate it, wanteth a great facultie. But there is a ſecond cauſe, which is no inabilitie, but a reiection
vpon choiſe and iudgement. For the honeſt and iuſt
bounds of obſeruation, by one perſon vpon ano-

ther,

ther, extend no further, but to vnderſtand him ſufficiently, whereby not to giue him offence, or wher-
by to be able to giue him faithfull Counſel, or wher-
by to ſtand vpon reaſonable guard and caution in reſpeƈt of a mans ſelfe: But to be ſpeculatiue into a-
nother man, to the end to know how to worke him,
or winde him, or gouerne him, proceedeth from
a heart that is double and clouen, and not entire and
ingenuous; which as in friendſhip it is want of integritie, ſo towards Princes or Superiors, is want of
dutie. For the cuſtome of the Leuant, which is, that
ſubieƈts doe forbeare to gaze or fixe their eyes vpon
Princes, is in the outward Ceremonie barbarous;
but the morall is good: For men ought not by cunning and bent obſeruations to pierce and penetrate
into the hearts of Kings, which the ſcripture hath
declared to be inſcrutable.

There is yet another fault (with which I will cor-
clude this part) which is often noted in learned men,
that they doe many times fayle to obſerue decencie,
and diſcretion in their behauiour and carriage, and
commit errors in ſmall and ordinarie points of aƈti-
on; ſo as the vulgar ſort of Capacities, doe make a
Iudgement of them in greater matters, by that
which they finde wanting in them, in ſmaller. But
this conſequence doth oft deceiue men, for which,
I doe referre them ouer to that which was ſayd by
Themiſtocles arrogantly, and vnciuily, being applyed
to himſelfe out of his owne mouth, but being applyed to the generall ſtate of this queſtion pertinent-

ly

ly and iuftly; when being inuited to touch a Lute,
he fayd: *He could not fiddle, but he could make a fmall
Towne, a great ftate.* So no doubt, many may be well
feene in the paffages of gouernement and policie,
which are to feeke in little, and punctuall occafions;
I referre them alfo to that, which *Plato* fayd of his
Maifter *Socrates*, whom he compared to the Gally-
pots of Apothecaries, which on the out fide had
Apes and Owles, and Antiques, but contained with
in foueraigne and precious liquors, and confections;
acknowledging that to an externall report, he was
not without fuperficiall leuities, and deformities; but
was inwardly replenifhed with excellent vertues
and powers. And fo much touching the point of
menners of learned men.

But in the meane time, I haue no purpofe to giue
allowance to fome conditions and courfes bafe, and
vnworthy, wherein diuers Profeffors of learning,
haue wronged themfelues, and gone too farre; fuch
as were thofe Trencher Philofophers, which in the
later age of the Romane State, were vfually in the
houfes of great perfons, being little better than folemne Parafites; of which kinde, *Lucian* maketh a
merrie defcription of the Philofopher that the great
Ladie tooke to ride with her in her Coach, and
would needs haue him carie her little Dogge which
he doing officioufly, and yet vncomely, the Page
fcoffed. and fayd: *That he doubted, the Philofopher of
a Stoike, would turne to be a Cynike.* But aboue all the
reft, the groffe and palpable flatterie, whereunto

E   many

many (not vnlearned) haue abbafed & abufed their
wits and pens, turning (as *Du Bartas* faith,) *Hecuba*
into *Helena*, and *Fauftina* into *Lucretia*, hath moft di-
minifhed the price and eftimation of Learning Neither is the morall dedications of Bookes and Writings, as to Patrons to bee commended: for that
Bookes (fuch as are worthy the name of Bookes)
ought to haue no Patrons, but Truth and Reafon:
And the ancient cuftome was, to dedicate them only to priuate and equall friendes, or to intitle the
Bookes with their Names, or if to Kings and great
perfons, it was to fome fuch as the argument of the
Booke was fit and proper for; but thefe and the like
Courfes may deferue rather reprehenfion, than defence.

Not that I can taxe or condemne the morigera-
tion or application of learned men to men in fortune. For the anfwere was good that *Diogenes* made
to one that asked him in mockerie, *How it came to
paffe that Philofophers were the followers of rich men, and
not rich men of Philofophers ?* He anfwered foberly,
and yet fharpely; *Becaufe the one fort knew what they
had need of, & the other did not;* And of the like nature
was the anfwere which *Ariftippus* made, when hauing a petition to *Dionifius*, and no eare giuen to
him, he fell downe at his feete, wheupon *Dionifius*
ftayed, and gaue him the hearing, and graunted it,
and afterward fome perfon tender on the behalfe
Philofophie, reprooued *Ariftippus*, that he would
o.fer the Profefsion of Philofophie fuch an indignitie,

tie, as for a priuat Suit to fall at a Tyrants feet: But he
anſwered; *It was not his fault, but it was the fault of
Dioniſius, that had his eares in his feete.* Neither was it
accounted weakeneſſe, but diſcretion in him that
would not diſpute his beſt with *Adrianus Cæſar;* ex-
cuſing himſelfe, *That it was reaſon to yeeld to him, that
commaunded thirtie Legions.* Theſe and the like applications and ſtooping to points of neceſſitie and
conuenience cannot bee diſallowed : for though
they may haue ſome outward baſeneſſe ; yet in a
Iudgement truely made, they are to bee accounted
ſubmiſsions to the occaſion, and not to the perſon.

Now I proceede to thoſe errours and vanities,
which haue interueyned amongſt the ſtudies themſelues of the learned; which is that which is principall and proper to the preſent argument, wherein
my purpoſe is not to make a iuſtification of the errors, but by a cenſure and ſeparation of the errors,
to make a iuſtificatiō of that which is good & ſound;
and to deliuer that from the aſperſion of the other.
For we ſee, that it is the manner of men, to ſcanda-
lize and depraue that which retaineth the ſtate, and
vertue, by taking aduantage vpon that which is corrupt and degenerate; as the Heathens in the primitiue Church vſed to blemiſh and taynt the Chriſtians with the faults and corruptions of Heretiques :
But neuertheleſſe, I haue no meaning at this time to
make any exact animaduerſion of the errors and
impediments in matters of learning, which are
more ſecret and remote from vulgar opinion ; but

onely

onely to ſpeake vnto ſuch as doe fall vnder, or neere
vnto, a popular obſeruation.

There be therfore chiefely three vanities in Studies, whereby learning hath been moſt traduced: For
thoſe things we do eſteeme vaine, which are either
falſe or friuolous, thoſe which either haue no truth,
or no vſe: & thoſe perſons we eſteem vain, which are
either credulous or curious, & curioſitie is either in
mater or words; ſo that in reaſon, as wel as in exper-
ence, there fal out to be theſe 3. diſtēpers (as I may
tearm thē) of learning; The firſt fantaſtical learning:
The ſecond contentious learning, & the laſt delicate
learning, vaine Imaginations, vaine Altercations,
& vain affectatiōs: & with the laſt I wil begin, *Martin
Luther* conducted (no doubt) by an higher prouidence, but in diſcourſe of reaſon, finding what a
Prouince he had vndertaken againſt the Biſhop of
Rome, and the degenerate traditions of the Church,
and finding his owne ſolitude, being no waies ayded
by the opinions of his owne time, was enforced to
awake all Antiquitie, and to call former times to his
ſuccors, to make a partie againſt the preſent time:
ſo that the ancient Authors, both in Diuinitie and
in Humanitie, which had long time ſlept in Libraries, began generally to be read and reuolued. This
by conſequence, did draw on a neceſsitie of a more
exquiſite trauaile in the languages originall, wherin
thoſe Authors did write: For the better vnderſtading
of thoſe Authors, and the better aduantage of preſſing and applying their words: And thereof grew
again,

againe, a delight in their manner of Stile and Phrase, and an admiration of that kinde of writing; which was much furthered & precipitated by the enmity & opposition, that the propounders of those (primitiue, but seeming new opinions) had against the Schoole-men: who were generally of the contrarie part: and whose Writings were altogether in a differing stile and fourme, taking libertie to coyne, and frame new tearms of Art, to expresse their own fence, and to auoide circuite of speech, without regard to the purenesse, pleasantnesse, and (as I may call it) lawfulnesse of the Phrase or word: And againe, because the great labour that then was with the people (of whome the Pharisees were wont to say: *Execrabilis ista turba quæ non nouit legem*) for the winning and perswading of them, there grewe of necessitie in cheefe price, and request, eloquence and varietie of discourse, as the fittest and forcibliest accesse into the capacitie of the vulgar sort: so that these foure causes concurring, the admiration of ancient Authors the hate of the Schoole-men, the exact studie of Languages: and the efficacie of Preaching did bring in an affectionate studie of eloquence, and copie of speech, which then began to flourish. This grew speedily to an excesse: for men began to hunt more after wordes, than matter, and more after the choisenesse of the Phrase, and the round and cleane composition of the sentence, and the sweet falling of the clauses, and the varying and illustration of their workes with tropes and figures:

then

then after the weight of matter, worth of fubie&t,
foundneffe of argument, life of inuention,or depth
of iudgement. Then grew the flowing, and wa-
trie vaine of *Oforius* the Portugall Bifhop, to be in
price: then did *Sturmius* fpend fuch infinite, and
curious paines vpon *Cicero* the Orator, and *Hermo-*
genes the Rhetorican, befides his owne Bookes of
Periods, and imitation, and the like: Then did *Car*
of *Cambridge*, and *Afcham* with their Lectures and
Writings, almoft diefie *Cicero* and *Demofthenes*, and
allure, all young men that were ftudious vnto that
delicate and pollifhed kinde of learning. Then did
Erafmus take occafion to make the fcoffing Eccho;
Decem annos confumpfi in legendo Cicerone: and the
Eccho anfwered in Greeke, *Oue; Afine.* Then grew
the learning of the Schoole-men to be vtterly defpifed as barbarous. In fumme, the whole inclination
and bent of thofe times, was rather towards copie,
than weight.

Here therefore, the firft diftemper of learning,
when men ftudie words, and not matter : whereof
though I haue reprefented an example of late times:
yet it hath beene, and will be *Secundum maius & mi-*
nus in all time. And how is it pofsible, but this fhould
haue an operation to difcredite learning, euen with
vulgar capacities, when they fee learned mens
workes like the firft Letter of a Patent, or limmed
Booke: which though it hath large flourifhes, yet it
is but a Letter. It feemes to me that *Pigmalions* fren-
zie is a good embleme or portraiture of this vanitie:

for wordes are but the Images of matter, and except they haue life of reafon and inuention: to fall in loue with them, is all one, as to fall in loue with a Picture.

But yet notwithftanding, it is a thing not haftily to be condemned, to cloath and adorne the the obfcuritie, euen of Philofophie it felfe, with fenfible and plaufible elocution. For hereof we haue great examples in *Xenophon*, *Cicero*, *Seneca*, *Plutarch*, and of *Plato* alfo in fome degree, and hereof likewife there is great vfe: For furely, to the feuere inquifition of truth, and the deepe progreffe into Philofophie, it is fome hindrance; becaufe it is too early fatisfactorie to the minde of man, and quencheth the defire of further fearch, before we come to a iuft periode. But then if a man be to haue any vfe of fuch knowledge in ciuile occafions, of conference, counfell, perfwafion, difcourfe, or the like: Then fhall he finde it prepared to his hands in thofe Authors, which write in that manner. But the exceffe of this is fo iuftly contemptible, that as *Hercules*, when hee faw the Image of *Adonis*, *Venus Migmon* in a Temple, fayd in difdaine, *Nil facri es*. So there is none of *Hercules* followers in learning, that is, the more feuere, and laborious fort of Enquirers into truth, but will defpife thofe delicacies and affectations, as indeede capable of no diuineffe. And thus much of the firft difeafe or diftemper of learning.

The fecond which followeth is in nature worfe then than the former: for as fubftance of matter is

better

better than beautie of words: so contrariwise vaine matter is worse, than vaine words: wherein it seemeth the reprehension of Saint *Paule*, was not onely proper for those times, but prophetical for the times following, and not only respectiue to Diuinitie, but extensiue to all knowledge. *Deuita prophanas vocum nouitates & oppositiones falsi nominis scientiæ.* For he assigneth two Markes and Badges of suspected and falsified Science; The one, the noueltie and strangenesse of tearmes; the other, the strictnesse of positions, which of necessitie doth induce oppositions, and so questions and altercations. Surely, like as many substances in nature which are solide, do putrifie and corrupt into wormes: So it is the propertie of good and sound knowledge, to putrifie and dissolue into a number of subtile idle, vnholesome, and (as I may tearme them) vermiculate questions; which haue indeede a kinde of quicknesse, and life of spirite, but no soundnesse of matter, or goodnesse of qualitie. This kinde of degenerate learning did chiefely raigne amongst the Schoole men, who hauing sharpe and stronge wits, and aboundance of leasure, and smal varietie of reading; but their wits being shut vp in the Cels of a few Authors (chiefely *Aristotle* their Dictator) as their persons were shut vp in the Cells of Monasteries and Colledges, and knowing little Historie, either of Nature or time, did out of no great quantitie of matter, and infinite agitation of wit, spin out vnto vs those laboriouswebbes of Learning which are extant in their Bookes.

Bookes. For the wit and minde of man, if it worke vpon matter, which is the contēplation of the creatures of God worketh according to the ſtuffe, and is limited thereby; but if it worke vpon it ſelfe, as the Spider worketh his webbe, then it is endleſſe, and and brings forth indeed Copwebs of learning, admirable for the fineſſe of thread and worke, but of no ſubſtance or profite.

This ſame vnprofitable ſubtilitie or curioſitie is of two ſorts: either in the ſubieƈt it ſelfe that they handle, when it is a fruitleſſe ſpeculation or controuerſie; (whereof there are no ſmall number both in Diuinity & Philoſophie) or in the maner or method of handling of a knowledge; which amongſt them was this; vpon euerie particular poſition or aſſertion to frame obieƈtions, and to thoſe obieƈtious, ſolutions: which ſolutions were for the moſt part not confutations, but diſtinƈtions: whereas indeed the ſtrength of all Sciences, is as the ſtrength of the old mans faggot in the bond. For the harmonie of a ſcience ſupporting each part the other, is and ought to be the true and briefe confutation and ſuppreſſion of all the ſmaller ſort of obieƈtions: but on the other ſide, if you take out euerie Axiome, as the ſtickes of the faggot one by one, you may quarrell with them, and bend them and breake them at your pleaſure: ſo that as was ſayd of *Sexaa: Verborum minutijs rerum frangit pondera*: So a man may truely ſay of the Schoole men *Quaſtionum minutijs Scientiarum frangunt ſoliditatem*. For were it not better for a man in a faire roome, to fet vp one great light, or braunching candlefticke of lights, than to goe about with a fmall watch candle into euerie corner? and fuch is their methode, that refts not fo much vppon euidence of truth prooued by arguments, authorities, fimilitudes, examples; as vpon particular confutations and folutions of euerie fcruple, cauillation & obiection: breeding for the moft part one queftiõ as faft as it folueth another; euẽ as in the former refemblance, when you carry the light into one corner, you darken the reft: fo that the Fable and fiction of *Scylla* feemeth to be a liuely Image of this kinde of Philofophie or knowledge, which was transformed into a comely Virgine for the vpper parts; but then, *Candida fuccinctam, latrantibus inguina monftris* : So the Generalities of the Schoolemen are for a while good and proportionable; but then when you defcend into their diftinctions and decifions, in ftead of a fruitfull wombe, for the vfe and benefite of mans life; they end in monftrous altercations and barking queftions. So as it is not poffible but this qualitie of knowledge muft fall vnder popular contempt, the people being apt to contemne truth vpõ occafion of Controuerfies & altercations, and to thinke they are all out of their way which neuer meete, and when they fee fuch digladiation about fubtilties, and matter of no vfe nor moment, they eafily fall vpon that iudgement of *Dionyfius* of *Siracufa, Verba ifta funt fenum otioforum.*

Notwithftanding certaine it is, that if thofe
fchoole-

Schoole men to their great thirst of truth, and vn-wearied trauaile of wit, had ioyned varietie and vni-uersalitie of reading and contemplation, they had prooued excellent Lights, to the great aduance-ment of all learning and knowledge: but as they are, they are great vndertakers indeed, and fierce with darke keeping. But as in the inquirie of the diuine truth, their pride enclined to leaue the Oracle of Gods word, and to vanish in the mixture of their owne inuentions : so in the inquisition of Nature, they euer left the Oracle of Gods works, and ado-red the deceiuing and deformed Images, which the vnequall mirrour of their owne minds, or a few re-ceiued Authors or principles, did represent vnto them. And thus much for the second disease of lear-ning.

For the third vice or disease of Learning, which concerneth deceit or vntruth, it is of all the rest the fowlest ; as that which doth destroy the essentiall fourme of knowledge ; which is nothing but a re-presentation of truth ; for the truth of being, and the truth of knowing are one, differing no more than the direct beame, and the beame reflected. This vice therefore brauncheth it selfe into two sorts ; delight in deceiuing, and aptnesse to be deceiued, imposture and Credulitie: which although they ap-peare to be of a diuers nature, the one seeming to proceede of cunning, and the other of simplicitie ; yet certainely, they doe for the most part concurre : for as the verse noteth.

Of the aduancement of learning,

Percontatorem fugito, nam Garrulus idem est:

An inquisitiue man is a pratler: so vpon the like reason, a credulous man is a deceiuer: as we see it in fame, that hee that will easily beleeue rumors, will as easily augment rumors, and adde somewhat to them of his owne, which *Tacitus* wisely noteth, when he sayth: *Fingunt simul creduntq;* so great an affinitie hath fiction and beleefe.

This facilitie of credite, and accepting or admitting thinges weakely authorized or warranted, is of two kindes, according to the subiect: For it is either a beleefe of Historie, (as the Lawyers speake, matter of fact:) or else of matter of art and opinion; As to the former, wee see the experience and inconuenience of this errour in ecclesiasticall Historie, which hath too easily receiued and regiſtred reports and narrations of Miracles wrought by Martyrs, Hermits, or Monkes of the desert, and other holy men; and there Reliques, Shrines, Chappels, and Images: which though they had a paſſage for time, by the ignorance of the people, the superstitious simplicitie of some, and the politique tolleration of others, holding them but as diuine poesies: yet after a periode of time, when the mist began to cleare vp, they grew to be esteemed, but as old wiues fables, impoſtures of the Cleargie illusions of spirits, and badges of Antichrist, to the great scandall and detriment of Religion.

So in naturall Historie, wee see there hath not
beene

beene that choife and iudgement vfed, as ought
to haue beene, as may appeare in the writings of
Plinius, Cardanus, Albertus, and diuers of the Ara-
bians, being fraught with much fabulous matter,
a great part, not onely vntryed, but notorioufly vntrue, to the great derogation of the credite of naturall Philofophie, with the graue and fober kinde
of wits; wherein the wifedome and integritie of *A-
riftotle* is worthy to be obferued, that hauing made
fo diligent and exquifite a Hiftorie of liuing creatures, hath mingled it fparingly with any vaine
or fayned matter, and yet on thother fake, hath caft
all prodigious Narrations, which he thought worthy the recording into one Booke: excellently difcerning that matter of manifeft truth, fuch wherevpon obferuation and rule was to bee built, was
not to bee mingled or weakened with matter of
doubtfull credite: and yet againe that rarities
and reports, that feeme vncredible, are not to
be fuppreffed or denyed to the memorie of men.

And as for the facilitie of credite which is yeel-
ded to Arts & opinions, it is likewife of two kinds,
either when too much beleefe is attributed to the
Arts themfelues, or to certaine Authors in any Art.
The Sciences themfelues which haue had better
intelligence and confederacie with the imagination
of man, than with his reafon, are three in number;
Aftrologie, Naturall Magicke, and *Alcumy*: of which
Sciences neuertheleffe the ends or pretences are
noble. For Aftrologie pretendeth to difcouer that

correspondence or concatenation, which is betweene the superiour Globe and the inferiour. Naturall Magicke pretendeth to cal & reduce natural
Philosophie from variety of speculations to the magnitude of works; And *Alcumy* pretendeth to make
separation of all the vnlike parts of bodies, which
in mixtures of nature are incorporate. But the deriuations and prosecutions to these ends, both in the
theories, and in the practises are full of Errour and
vanitie; which the great Professors themselues haue
sought to vaile ouer and conceale by enigmaticall
writings, and referring themselues to auricular traditions, and such other deuises, to saue the credite
of Impostures; and yet surely to *Alcumy* this right
is due, that it may be compared to the Husband man
whereof *Æsope* makes the Fable; that when he died, told his Sonnes, that he had left vnto them gold,
buried vnder ground in his Vineyard; and they dig-
ged ouer all the ground, and gold they found none,
but by reason of their stirring and digging the mold
about, the rootes of their Vines, they had a great
Vintage the yeare following: so assuredly the search
and stirre to make gold hath brought to light a great
number of good and fruitfull inuentions and experiments, as well for the disclosing of Nature; as for
the vse of mans life.

 And as for the ouermuch credite that hath beene
giuen vnto Authors in Sciences, in making them
Dictators, that their wordes should stand, and not
Counsels to giue aduise; the dammage is infinite that
Sciences

Sciences haue receiued thereby, as the principall cause that hath kept them lowe, at a stay without groweth or aduancement. For hence it hath comen, that in arts Mechanicall, the first deuiser coms shortest, and time addeth and perfecteth: but in Sciences the first Author goeth furthest, and time leeseth and corrupteth. So we see, Artillerie, sayling, printing, and the like, were grossely managed at the first and by time accommodated and refined: but contrarywise the Philosophies and Sciences of *Aristotle, Plato, Democritus, Hypocrates, Euclides, Archimedes,* of most vigor at the first, and by time degenerate and imbased, whereof the reason is no other, but that in the former many wits and industries haue haue contributed in one; and in the later many wits and industries haue ben spent about the wit of some one; whom many times they haue rather depraued than illustrated. For as water will not ascend higher, than the leuell of the first spring head, from whence it descendeth: so knowledge deriued from *Aristotle,* and exempted from libertie of examination, will not rise againe higher, than the knowledge of *Aristotle.* And therfore although the position be good: *Oportet discentem credere:* yet it must bee coupled with this, *Oportet edoctum iudicare:* for Disciples doe owe vnto Maisters onely a temporarie beleefe, and a suspension of their owne iudgement, till they be fully instructed. and not an absolute resignation, or perpetuall captiuitie: and therefore to conclude this point, I will say no more, but; so let great Authors

F 4   haue

haue theire due, as time which is the Author of Authors be not depriued of his due, which is furder and furder to difcouer truth. Thus haue I gone o-uer thefe three difeaffes of learning, befides the which there are fome other rather peccant humors, then fourmed difeafe s, which neuertheles are not fo fecret and intrinfike, but that they fall vnder a po-pular obferuation and traducement; and therefore are not to be paffed ouer.

The firft of thefe is the extreame affecting of two extreamities; The one Antiquity, The other No-uelty; wherein it feemeth the children of time doe take after the nature and mallice of the father. For as he deuowreth his children; fo one of them feek-eth to deuoure and fuppreffe the other; while An-tiquity enuieth there fhould be new additions, and ⁊ 'ouelty; cannot be content to add, but it muft de-face; Surely the aduife of the Prophet is the true di-rection in this matter, *State fuper vias antiquas, & vi-dete quanam fit via recta & bona, & ambulate in ea.* Antiquity deferueth that reuerēce, that men fhould make a ftand thereupon, and difcouer what is the beft way, but when the difcouery is well taken then to make progreffion. And to fpeake truly, *Antiqui-ta feculi Iuuentus Mundi.* Thefe times are the ancient times when the world is ancient, & not thofe which we count antient *Ordine retrogrado,* by a computa-cion backward from our felues.

Another Error induced by the former is a diftruft that any thing fhould bee now to bee found out
which

which the world ſhould haue miſſed and paſſed
ouer ſo long time, as if the ſame obiection were to
be made to time, that *Lucian* maketh to *Iupiter*, and
other the heathen Gods, of which he woondreth,
that they begot ſo many Children in old time, and
begot none in his time, and asketh whether they
were become ſeptuagenarie, or whether the lawe
Pappia made againſt old mens mariages had reſtray-
ned them. So it ſeemeth men doubt, leaſt time is become paſt children and generation; wherein contrary wiſe, we ſee commonly the leuitie and vncon-
ſtancie of mens iudgements, which till a matter bee
done, wonder that it can be done; and aſſoone as it is
done, woonder againe that it was no ſooner done,
as we ſee in the expedition of *Alexander* into *Aſia*,
which at firſt was preiudged as a vaſt and impoſsible
enterprize; and yet afterwards it pleaſeth *Liuye* to
make no more of it, than this, *Nil aliud quam bene
auſus vana contemnere.* And the ſame happened to
Columbus in the weſterne Nauigation. But in intellectuall matters, it is much more common; as may
be ſeen in moſt of the propoſitions of *Euclyde*, which
till they bee demonſtrate, they ſeeme ſtrange to our
aſſent; but being demonſtrate, our mind accepteth
of them by a kind of relation (as the Lawyers ſpeak)
as if we had knowne them before.

Another Errour that hath alſo ſome affinitie
with the former is a conceit that of former opinions
or ſects after varietie and examination, the beſt hath
ſtill preuailed; and ſuppreſſed the reſt; So as if a
G   man

man fhould beginne the labour of a newe fearch, hee were but like to light vppon fomewhat formerly reiected ; and by reiection, brought into obliuion; as if the multitude, or the wifeft for the multitudes fake, were not readie to giue paffage, rather to that which is popular and fuperficiall, than to that which is fubftantiall and profound; for the truth is, that time feemeth to be of the nature of a Riuer, or ftreame , which carryeth downe to vs that which is light and blowne vp; and finketh and drowneth that which is weightie and folide.

Another Errour of a diuerfe nature from all the former, is the ouer-early and peremptorie reduction of knowledge into Arts and Methodes: from which time, commonly Sciences receiue fmall or no augmentation. But as young men, when they knit and fhape perfectly, doe feldome grow to a further ftature : fo knowledge, while it is in Aphorifmes and obferuations, it is in groweth ; but when it once is comprehended in exact Methodes; it may perchance be further pollifhed and illuftrate, and accommodated for vfe and practife ; but it encreafeth no more in bulke and fubftance.

Another Errour which doth fucceed that which we laft mentioned, is, that after the diftribution of particular Arts and Sciences, men haue abandoned vniuerfalitie, or *Philofophia prima;* which cannot but ceafe, and ftoppe all progresfion. For no perfect difcouerie can bee made vppon a flatte, or a leuell.
Neither

Neither is it possible to discouer the more remote, and deeper parts of any Science, if you stand but vpon the leuell of the same Science, and ascend not to a higher Science.

Another Error hath proceeded from too great a reuerence, and a kinde of adoration of the minde and vnderstanding of man: by meanes whereof, men haue withdrawne themselues too much from the contemplation of Nature, and the obseruations of experience: and haue tumbled vp and downe in their owne reason and conceits: vpon these Intelle-ctuallists, which are notwithstanding commonly taken for the most sublime and diuine Philosophers; *Heraclitus* gaue a iust censure, saying : *Men sought truth in their owne little worlds, and not in the great and common world :* for they disdaine to spell, and so by degrees to read in the volume of Gods works, and contrarywise by continuall meditation and agitation of wit, doe vrge, and as it were inuocate their owne spirits, to diuine, and giue Oracles vnto them, whereby they are deseruedly deluded.

Another Error that hath some connexion with this later, is, that men haue vsed to infect their meditations, opinions, and doctrines with some conceits which they haue most admired, or some Sciences which they haue most applyed; and giuen all things else a tincture according to them, vtterly vntrue and vnproper. So hath *Plato* intermingled his Philosophie with Theologie, and *Aristotle* with Logicke, and the second Schoole of *Plato,*

Proclus, and the reſt, with the Mathematiques. For
theſe were the Arts which had a kinde of *Primo ge-
niture* with them ſeuerally. So haue the Alchymiſts
made a Philoſophie out of a few experiments of the
Furnace; and *Gilbertus* our Countrey man hath
made a Philoſophie out of the obſeruations of a
Loadſtone. So *Cicero*, when reciting the ſeuerall o-
pinions of the nature of the ſoule, he found a Muſitian, that held the ſoule was but a harmonie, ſayth
pleaſantly: *Hic ab arte ſua non receſſit,&c*.But of theſe
conceits *Ariſtotle* ſpeaketh ſeriouſly and wiſely,
when he ſayth: *Qui reſpiciunt ad pauca de facili pro-
nuntiant*.

Another Errour is an impatience of doubt, and
haſt to aſſertion without due and mature ſuſpention
of iudgement. For the two wayes of contemplation are not vnlike the two wayes of action, commonly ſpoken of by the Ancients. The one plain and
ſmooth in the beginning, and in the end impaſſable:
the other rough and troubleſome in the entrance,
but after a while faire and euen, ſo it is in cotempla-
tion, if a man will begin with certainties, hee ſhall
end in doubts; but if he will be content to beginne
with doubts, he ſhall end in certainties.

Another Error is in the manner of the tradition
and deliuerie of knowledge, which is for the moſt
part Magiſtrall and peremptorie; and not ingenuous and faithfull, in a ſort, as may be ſooneſt beleeued; and not eaſileſt examined. It is true, that in
compendious Treatiſes for practiſe, that fourme is
not

not to bee diſallowed. But in the true handling of knowledge, men ought not to fall either on the one ſide into the veyne of *Velleius* the Epicurean : *Nil tam metuens, quam ne dubitare aliqua de re videretur;* Nor on the other ſide into *Socrates* his irronicall doubting of all things, but to propound things ſincerely, with more or leſſe aſſeueration: as they ſtand in a mans owne iudgement, prooued more or leſſe.

Other Errors there are in the ſcope that men propound to themſelues, whereunto they bend their endeauours : for whereas the more conſtant and deuote kind of Profeſſors of any ſcience ought to propound to themſelues, to make ſome additions to their Science; they conuert their labours to aſpire to certaine ſecond Prizes; as to be a profound Interpreter or Cōmeater; to be a ſharpe Champion or Defender; to be a methodicall Compounder or abridger ; and ſo the Patrimonie of knowledge commeth to be ſometimes improoued but ſeldome augmented.

But the greateſt Error of all the reſt, is the miſtaking or miſplacing of the laſt or furtheſt end of knowledge : for men haue entred into a deſire of Learning and knowledge, ſometimes vpon a naturall curioſitie, and inquiſitiue appetite ; ſometimes to entertaine their mindes with varietie and delight; ſometimes for ornament and reputation; and ſometimes to inable them to victorie of wit and contradiction, and moſt times for lukar and profesſion, and ſeldome ſincerely to giue a true account of their

guiſt.

guift of reaſon, to the benefite and vſe of men : As
if there were ſought in knowledge a Cowch, wher-
vpon to reſt a ſearching and reſtleſſe ſpirite; or a tar-
raſſe for a wandring and variable minde, to walke
vp and downe with a faire proſpect ; or a Tower of
State for a proude minde to raiſe it ſelfe vpon; or a
Fort or commaunding ground for ſtrife and contention, or a Shoppe for profite or ſale; and not a
rich Store-houſe for the glorie of the Creator, and
the reliefe of Mans eſtate. But this is that, which
will indeed dignifie and exalt knowledge ; if contemplation and action may be more neerely and
ſtraightly conioyned and vnited together , than
they haue beene; a Coniunction like vnto that of
the two higheſt Planets, *Saturne* the Planet of reſt
and contemplation ; and *Iupiter* the Planet of ciuile ſocietie and action . Howbeit, I doe not
meane when I ſpeake of vſe and action, that end
before mentioned of the applying of knowledge
to luker and profeſſion ; For I am not ignorant
howe much that diuerteth and interrupteth the
proſecution and aduauncement of knowledge;
like vnto the goulden ball throwne before *Ata-*
lanta, which while ſhee goeth aſide, and ſtoo-
peth to take vp, the race is hindred,

Declinat curſus, aurumque volubile tollit :

Neither is my meaning as was ſpoken of *Socrates*, to
call Philoſophy down from heauē to conuerſe vpon
the earth, that is, to leaue natural Philoſophy aſide, &
to applye knowledge onely to manners, and policie.

But

But as both heauen and earth doe conſpire and contribute to the vſe and benefite of man: So the end ought to bee from both Philoſophies, to ſe-parate and reiect vaine ſpeculations, and whatſo-euer is emptie and voide, and to preſerue and augment whatſoeuer is ſolide and fruitfull: that knowledge may not bee as a Curtezan for pleaſure, & vanitie only, or as a bond-woman to acquire and gaine to her Maſters vſe, but as a Spouſe, for genera-tion, fruit, and comfort.

Thus haue I deſcribed and opened as by a kinde of diſſection, thoſe peccant humors (the principall of them) which hath not onely giuen impediment to the proficience of Learning, but haue giuen alſo occaſion, to the traducement thereof: wherein if I haue beene too plaine, it muſt bee remembred; *Fidelia vulnera amantis, ſed doloſa oſcula malignantis.* This I thinke I haue gained, that I ought to bee the better beleeued, in that which I ſhall ſay pertayning to commendation: becauſe I haue proceeded ſo freely, in that which concerneth cenſure. And yet I haue no purpoſe to enter into a laudatiue of Lear-ning, or to make a Hymne to the Muſes (though I am of opinion, that it is long ſince their Rites were duely celebrated) but my intent is without varniſh or amplification, iuſtly to weigh the dignitie of knowledge in the ballance with other things, and to take the true value thereof by teſtimonies and ar-guments diuine, and humane.

Firſt therefore, let vs ſeeke the dignitie of know-

ledge

ledge in the Arch-tipe or firft plat forme, which is
is in the attributes and acts of God, as farre as they
are reuealed to man, and may be obferued with fobrietie, wherein we may not feeke it bythe name
of Learning, for all learning is knowledge acqui-
red, and all knowledge in God is originall. And
therefore we muft looke for it by another name, that
of wifedome or fapience, as the fcriptures call it.

It is fo then, that in the worke of the Creation, we
fee a double emanation of vertue frō God : the one
referring more properly to power, the other to wifedome, the one expreffed in making the fubliftence of
the mater, & the other in difpofing the beauty of the
fourme. This being fuppofed, it is to bee obferued,
that for any thing which appeareth in the hiftorie of
the Creation, the confufed Maffe, and matter of heauen and earth was made in a moment, and the order
and difpofition of that *Chaos* or Maffe, was the work
of fixe dayes, fuch a note of difference it pleafed
God to put vppon the workes of power, and the
workes of wifedome: wherewith concurreth that
in the former, it is not fette downe, that God fayd,
Let there be Heauen and Earth, as it is fet downe of
the workes following, but actually, that God made
Heauen and earth: the one carrying the ftile of a
Manufacture, and the other of a lawe, decree, or
Councell.

To proceede to that which is next in order from
God to fpirits : we finde as farre as credite is to bee
giuen to the celeftiall Hierarchye, of that fuppofed

Dionyfius,

Dionyſius the Senator of Athens the firſt place or
degree is giuen to the Angels of loue, which are
tearmed *Seraphim*,the ſecond to the Angels of light,
which are tearmed *Cherubim*, and the third; and ſo
following places to thrones, principalities, and the
reſt, which are all Angels of power and miniſtry; ſo
as the Angels of knowledge and illumination, are
placed before the Angels of Office and domina-
tion.

To deſcend from ſpirits and intellectuall formes
to ſenſible and materiall fourmes, wee read the
firſt fourme that was created, was light, which
hath a relation and correſpondence in nature,and
corporall thinges, to knowledge in ſpirits and in-
corporall thinges.

So in the diſtribution of dayes,we ſee the day
wherin God did reſt, & contēplate his owne works,
was bleſſed aboue all the dayes, wherein he did effect and accompliſh them.

After the Creation was finiſhed, it is ſette
downe vnto vs, that man was placed in the
Garden to worke therein, which worke ſo appointed to him, could be no other than worke of
contemplation, that is, when the end of worke is
but for exerciſe and experiment, not for neceſsitie,
for there being then no reluctation of the creature,
nor ſweat of the browe, mans employment muſt
of conſequence haue ben matter of delight in the experiment and not matter of labor for the vſe.Againe
the firſt Acts which man perfourmed in Paradiſe,
H   conſiſted

confifted of the two fummarie parts of knowledge, the view of Creatures, and the impofition of names. As for the knowledge which induced the fall, it was, as was touched before, not the naturall knowledge of Creatures, but the morall knowledge of good and euill, wherein the fuppofition was, that Gods commaundements or prohibitions were not the originals of good and euill, but that they had other beginnings which man afpired to know, to the end, to make a totall defection from God, and to depend wholy vpon himfelfe.

To paffe on, in the firft euent or occurrence after the fall of Man; wee fee (as the Scriptures haue infinite Myfteries, not violating at all the truth of the Storie or letter) an Image of the two Eftates, the Contemplatiue ftate, and the actiue ftate, figured in the two perfons of *Abell* and *Cain,* and in the two fimpleft and moft primitiue Trades of life: that of the Shepheard (who by reafon of his leafure, reft in a place, and liuing in view of heauen, is a liuely Image of a contemplatiue life) and that of the hufbandman; where we fee againe, the fauour and election of God went to the Shepheard, and not to the tiller of the ground.

So in the age before the floud, the holy Records within thofe few memorials, which are there entred and regiftred, haue vouchfafed to mention, and honour the name of the Inuentors and Authors of Mufique, and works in mettall. In the age after the Floud, the firft great iudgement of God vppon the
ambition

ambition of man, was the confuſion of tongues; whereby the open Trade and intercourſe of Learning and knowledge, was chiefely imbarred.

To deſcend to *Moyſes* the Law-giuer, and Gods firſt penne; hee is adorned by the Scriptures with this addition, and commendation: *That he was ſeene in all the Learning of the Ægyptians*; which Nation we know was one of the moſt ancient Schooles of the world: for, ſo *Plato* brings in the Egyptian Prieſt, ſaying vnto *Solon*: *You Grecians are euer Children, you haue no knowledge of antiquitie, nor antiquitie of knowledge.* Take a view of the ceremoniall law of *Moyſes*; you ſhall find beſides the prefiguration of Chriſt, the badge or difference of the people of God, the excerciſe and impreſsion of obedience, and other diuine vſes and fruits thereof, that ſome of the moſt learned *Rabynes* haue trauailed profitably, and profoundly to obſerue, ſome of them a naturall, ſome of them a morall ſence, or reduction of many of the ceremonies and ordinances: As in the lawe of the Leprouſie, where it is ſayd: *If the whiteneſſe haue o-uerſpread the fleſh, the Patient may paſſe abroad for clean; But if there be any whole fleſh remayning, he is to be ſhut vp for vncleane:* One of them noteth a principle of nature, that putrefaction is more contagious before maturitie than after: And another noteth a poſition of morall Philoſophie, that men abandoned to vice, doe not ſo much corrupt manners, as thoſe that are halfe good, and halfe euill, ſo, in this and verie many other places in that lawe, there is to bee found be-

ſides

fides the Theologicall fence, much afperfion of Philofophie.

So likewife in that excellent Booke of *Iob*, if it be
reuolued with diligence, it will be found pregnant,
and fwelling with naturall Philofophie; as for example, Cofmographie, and the roundneffe of the
world : *Qui extendit aquilonem fuper vacuum, &
appendit terram fuper nihilum* : wherein the penfile-
neffe of the earth, the pole of the North, and
the finiteneffe, or conuexitie of Heauen are mani-
feftly touched. So againe matter of Aftronomie;
*Spiritus eius ornauit cœlos & obftetricante manu e-
ius eductus eft coluber tortuofus* : And in another
place, *Nunquid coniungere valebis micantes ftellas
pleyadas, aut gyrum arcturi poteris diſſipare?* where
the fixing of the ftarres, euer ftanding at equall
diftance, is with great elegancie noted : And in
another place, *Qui facit arcturum, & oriona, & hy-
adas, & interiora auſtri*, where againe hee takes
knowledge of the deprefsion of the Southerne pole,
calling it the fecrets of the South, becaufe the fou-
therne ftarres were in that climate vnfeene. Matter of generation, *Annon fi ut lac mulfifti me, & ficut
cafeum coàgulafti me, &c.* Matter of Mynerals, *Habet
argentum venarum fuarum principia : & auro locus eft
in quo conflatur, ferrum de terra tollitur, & lapis folutus
calore in æs vertitur* : and fo forwards in that Chapter.

So likewife in the perfon of *Salomon* the
King, wee fee the guift or endowment of wife-
dome and learning both in *Salomons* petition, and in
Gods

Gods affent thereunto preferred before all other terrene and temporall felicitie. By vertue of which grant or donatiue of God, *Salomon* became inabled, not onely to write thofe excellent Parables, or Aphorifmes concerning diuine and morall Philofophie; but alfo to compile a naturall Hiftorie of all verdor, from the Cedar vpon the Mountaine, to the moffe vppon the wall, (which is but a rudiment betweene putrefaction, and an hearbe) and alfo of all things, that breath or mooue. Nay the fame *Salomon* the King, although he excelled in the glorie of treafure and magnificent buildings of fhipping and Nauigation, of feruice and attendance, of fame and renowne, and the like; yet hee maketh no claime to any of thofe glories; but onely to the glorie of Inquifition of truth : for fo he fayth expreffely : *The glorie of God is to conceale a thing, But the glorie of the King is to find it out*, as if according to the innocent play of Children the diuine Maieftie tooke delight to hide his workes, to the end to haue them found out, and as if Kinges could not obtaine a greater honour, than to bee Gods playfellowes in that game, confidering the great commaundement of wits and meanes, whereby nothing needeth to be hidden from them.

Neither did the difpenfation of God varie in the times after our Sauiour came into the world; for our Sauiour himfelfe did firft fhew his power to fubdue ignorance, by his conference with the Priefts and Doctors of the lawe ; before he fhewed his power

to subdue nature by his miracles. And the comming
of the holy spirite, was chiefely figured and expressed in the similitude and guift of tongues; which are
but *Vehicula scientiæ.*

So in the election of those Instruments, which it
pleased God to vse for the plantation of the faith,
notwithstanding, that at the first hedid employ persons altogether vnlearned, otherwise than by inspiration, more euidently to declare his immediate
working, and to abbase all humane wisedome or
knowledge; yet neuerthelesse, that Counsell of his
was no sooner perfourmed, but in the next vicissitude and succession, he did send his diuine truth into the world, wayted on with other Learnings, as
with Seruants or Handmaides : For so we see *S*aint
Paule, who was only learned amongst the Apostles,
had his penne most vsed in the scriptures of the new
Testament.

*S*o againe, we finde that many of the ancient Bi-
shops and Father of the Church, were excellently
redde, & studied in all the learning of the Heathen,
insomuch, that the Edict of the Emperour *Iulianus*
(whereby it was interdicted vnto Christians to bee
admitted into *S*chooles, Lectures, or exercises of
learning) was esteemed and accounted a more pernitious engine and machination against the Christian faith; than were all the sanguinarie prosecutions
of his Predecessors; Neither could the emulation
and Iealousie of *Gregorie* the first of that name, Bi-
shop of *Rome,* euer obtaine the opinion of pietie or
deuotion:

deuotion: but contrarywiſe receiued the cenſure
of humour, malignitie, and puſillanimitie, euen a-
mongſt holy men: in that he deſigned to obliterate
and extinguiſh the memorie of Heathen antiquitie
and Authors. But contrarewiſe it was the Chriſtian Church, which amidſt the inundations of the
Scythians, on the one ſide from the Northweſt: and
the *Saracens* from the Eaſt, did preſerue in the ſacred
lappe and boſome thereof, the pretious Reliques, e-
uen of Heathen Learning, which otherwiſe had
beene extinguiſhed, as if no ſuch thing had euer
beene.

And wee ſee before our eyes, that in the age of
our ſelues, and our Fathers, when it pleaſed God to
call the Church of Rome to account, for their degenerate manners and ceremonies: and ſundrie doctrines, obnoxious, and framed to vphold the ſame
abuſes: At one and the ſame time, it was ordayned
by the diuine prouidence, that there ſhould attend
withall a renouation, and new ſpring of all other
knowledges: And on the other ſide, we ſee the Ieſuites, who partly in themſelues, and partly by the
emulation and prouocation of their example, haue
much quickned and ſtrengthned the ſtate of *Lear*-
ning: we ſee (I ſay) what notable ſeruice and repa-
ration they haue done to the Romane *Sea*.

Wherefore to conclude this part, let it bee obſerued, that there be two principall duties and ſeruices beſides ornament & illuſtration, which Philoſophie and humane learning doe perfourme to
H 4

faith

faith and Religion. The one, becaufe they are an effectuall inducement to the exaltation of the glory of God. For as the *Pfalmes*,and other Scriptures doe often inuite vs to confider, and magnifie the great and wonderfull workes of God : fo if we fhould reft onely in the contemplation of the exterior of them, as they firft offer themfelues to our fences; we fhould do a like iniurie vnto the Maieftie of God, as if wee fhould iudge or conftrue of the ftore of fome excellent Ieweller, by that onely which is fet out toward the ftreete in his fhoppe. The other, becaufe they minifter a finguler helpe and preferuatiue againft vnbeleefe and error ; For our Sauiour faith, *You erre not knowing the Scriptures,nor the power of God*: laying before vs two Bookes or volumes to ftudie, if we will be fecured from errour : firft the fcriptures, reuealing the will of Cod ; and then the creatures exprefsing his power ; whereof the later is a key vnto the former ; not onely opening our vnderftanding to conceiue the true fence of the fcriptures, by the generall notions of reafon and rules of fpeech ; but chiefely opening our beleefe, in drawing vs into a due meditation of the omnipotencie of God, which is chiefely figned and ingrauen vppon his workes. Thus much therefore for diuine teftimonie and euidence, concerning the true dignitie, and value of learning.

As for humane proofes, it is fo large a field, as in a difcourfe of this nature and breuitie, it is fit rather to vfe choife of thofe things, which we fhall produce,

duce, than to embrace the variety of them. First therfore in the degrees of humane honour amongst the
heathen, it was the highest, to obtain to a veneration
& adoration as a God. This vnto the Christians is as
the forbidden fruit. But we speake now separately of
humane testimonie; according to which, that which
the Grecians call Apotheosis, and the Latines, *Rela-*
tio inter diuos, was the supreame honour, which man
could attribute vnto man; specially when it was giuen, not by a formall Decree or Act of State, as it
was vsed amongst the Romane Emperours; but by
an inward assent and beleefe; which honour being
so high, had also a degree or middle tearme: for
there were reckoned aboue humane honours, honour heroycall and diuine: In the attribution, and
distribution of which honours; wee see Antiquitie
made this difference : that whereas founders and
vniters of States and Cities, Law-giuers, extirpers
of Tyrants, Fathers of the people, and other eminent persons in ciuile merite, were honoured but
with the titles of Worthies or Demy-Gods : such as
were *Hercules, Theseus, Minos, Romulus*, and the like:
on the other side, such as were Inuentors and Authors of new Arts, endowments, and commodities
towards mans life, were euer consecrated amongst
the Gods themselues, as was *Ceres, Bacchus, Mercurius,*
Apollo, and others, and iustly: for the merit of the former is confined within the circle of an age, or a nation : and is like fruitfull showers, which though
they be profitable and good : yet serue but for that

I   season,

feafon, and for a latitude of ground where they fall:
But the other is indeed like the benefits of Heauen,
which are permanent and vniuerfall. The former
againe is mixt with ftrife and perturbation; but the
later hath the true Caracter of diuine prefence; com-
in *aura leni*, without noife or agitation.

Neither is certainely that other merite of learning, in reprefsing the inconueniences which grow
from man to man; much inferiour to the former, of
relieuing the necefsities which arife from nature;
which merite was liuely fet forth by the Ancients
in that fayned relation of *Orpheus* Theater; where
all beafts and birds affembled; and forgetting their
feuerall appetites; fome of pray, fome of game, fome
of quarrell, ftood all fociably together liftening vnto
the ayres and accords of the Harpe; the found
whereof no fooner ceafed, or was drowned by
fome lowder noyfe; but euerie beaft returned to
his owne nature; wherein is aptly defcribed the nature and condition of men; who are full of fauage
and vnreclaymed defires; of profite, of luft, of reuenge; which as long as they giue eare to precepts,
to lawes, to religion, fweetely touched with eloquence and perfwafion of Bookes, of Sermons, of
haranges; fo long is focietie and peace maintained;
but if thefe inftruments bee filent; or that fedition
and tumult make them not audible; all thinges dif-
folue into Anarchie and Confufion.

But this appeareth more manifeftle, when Kings
themfelues, or perfons of authoritie vnder them or
other

other Gouernours in common wealthes, and popular Eſtates, are endued with Learning. For although
he might be thought partiall to his owne profeſsion,
that ſayd, *Then ſhould people and eſtates be happie, when
either Kings were Philoſophers, or Philoſophers Kings* :
yet ſo much is verified by experience; that vnder
learned Princes and Gouernours, there haue
been euer the beſt times ; for howſoeuer Kinges
may haue their imperfeƈtions in their paſsions and
Cuſtomes; yet if they be illuminate by learning, they
haue thoſe Notions of Religion, policie, and moralitie ; which doe preſerue them, and refraine them
from all ruinous and peremptory errors & exceſſes;
whiſpering euermore in their eares, when Counſellors and ſeruants ſtand mute, and ſilenƚ; and Senators, or Counſellours likewiſe, which bee learned,
doe proceede vpon more ſafe and ſubſtantiall principles ; then Counſellors which are onely men of
experience ; the one ſort keeping dangers a farre
off ; whereas the other diſcouer them not, till they
come neere hand : and then truſt to the agilitie of
their wit, to ward or auoide them.

Which felicitie of times, vnder learned Princes,
(to keepe ſtill the Lawe of breuitie, by vſing the
moſt eminent and ſeleƈted examples) doth beſt appeare in the age, which paſſed from the death of
Domitianus the Emperour, vntill the raigne of *Commodus* : comprehending a ſucceſsion of ſixe Sciences, all learned or ſinguler fauourers and Aduancers
of learning : which age for temporall reſpeƈts, was

I 2 the

the moſt happie and flouriſhing, that euer the Romane Empire, (which then was a modele of the
world) enioyed: a matter reuealed and prefigured
vnto *Domitian* in a Dreame, the night before he was
flaine; for hee thought there was growne behinde
vpon his fhoulders, a necke and a head of gould,
which came accordingly to paſſe, in thoſe golden
times which ſucceeded; of which Princes, we will
make ſome commemoration: wherein although the
matter will bee vulgar, and may be thought fitter
for a Declamation, then agreeable to a Treatiſe infol-
ded as this is; yet becauſe it is pertinent to the point
in hand, *Neque ſemper arcum tendit Apollo*, & to name
them onely were too naked and curforie, I will not
omit it altogether. The firſt was *Nerua*, the excellent temper of whoſe gouernement, is by a glaunce
in *Cornelius Tacitus* touched to the life: *Poſtquam di-
uus Nerua res olim inſociabiles miſcuiſſet, imperiū & li-
bertatem*: And in token of his learning, the laſt Act
of his ſhort raigne left to memorie, was a miſſiue to
his adopted ſonne *Traian*, proceeding vpon ſome
inward diſcontent, at the ingratitude of the times,
comprehended in a verfe of *Homers*,

Telis Phœbe, tuis, Lachrymas vlciſcere noſtras.

Traian, who ſucceeded, was for his perſon not
learned : But if wee will hearken to the ſpeech of
our Sauiour, that ſayth, *Hee that receiueth a Prophet in the name of a Prophet, ſhall haue a Prophets reward*, hee deſerueth to bee placed amongeſt the
moſt learned Princes : for there was not a greater
                                                    admirer

admirer of learning or Benefactor of Learning, a founder of famous Libraries, a perpetuall Aduancer of learned men to office, and a familiar conuerſer with learned Profeſſors and Preceptors, who were noted to haue then moſt credite in Court. On the other ſide, how much *Traians* vertue and gouernement was admired & renowned, ſurely no teſtimonie of graue and faithfull Hiſtory doth more liuely ſet forth, than that legend tale of *Gregorius Magnus*, Biſhop of Rome, who was noted for the extream enuy he bare towards all Heathen excellencie: and yet he is reported out of the loue and eſtimation of *Traians* morall vertues, to haue made vnto God, paſsionate and feruent prayers, for the deliuerie of his ſoule out of Hell: and to haue obtained it with a Caueat that he ſhould make no more ſuch petitions. In this Princes time alſo, the perſecutions againſt the Chriſtians receiued intermiſsion, vpon the certificate of *Plinius ſecundus*, a man of excellent learning, and by *Traian* aduanced

Adrian his ſucceſſor, was the moſt curious man that liued, and the moſt vniuerſal enquirer: inſomuch as it was noted for an errour in his mind: that he deſired to comprehend all thinges, and not to reſerue himſelfe for the worthyeſt thinges, falling into the like humour that was long before noted in *Phillip* of *Macedon*, who when hee would needs ouer rule and put downe an excellent Muſitian, in an argument touching Muſique, was well anſwered by him againe, *God forbid Sir* (ſaith hee)

that

that your fortune should be so bad, as to know these things better than I; It pleafed God likewife to vfe the curiofitie of this Emperour, as an inducement to the peace of his Church in thofe dayes : for hauing Chrift in veneration, not as a God or Sauiour, but as a wonder or noueltie: and hauing his picture in his Gallerie, matched with *Apollon.us* (with whom in his vaine imagination, he thought he had fome conformitie) yet it ferued the turne to allay the bitter hatred of thofe times againft the Chriftian name : fo as the Church had peace during his time, and for his gouernement ciuile, although he did not attaine to that of *Traians,* in glorie of Armes, or perfection of Iuftice : yet in deferuing of the weale of the Subiect, he did exceede him. For *Traiane* erected many famous monuments and buildings, infomuch as *Conftantine* the Great, in emulation was woont to call him *Parietaria,* Wall flower, becaufe his name was vppon fo many walles : but his buildings and workes were more of glorie and tryumph, than vfe and necefsitie: But *Adrian* fpent his whole Raigne, which was peaceable in a perambulation, or Suruey of the Romane Empire, giuing order and making afsignation, where he went for reedifying of Cities, Townes, and Forts decayed : and for cutting of Riuers, and ftreames : and for making Bridges and paffages, and for pollicing of Cities, and Commonalties, with new ordinances and conftitutions : and graunting new Franchifes and incorporations : fo that his whole time was a very reftauration of all the lapfes, lapſes and decayes of former times.

Antonius Pius, who ſucceeded him, was a Prince excellently learned ; and had rhe Patient and ſubtile witte of a Schoole man : inſomuch as in common ſpeech, (which leaues no vertue vntaxed) hee was called *Cymini Sector*, a caruer, or a diuider of Comine ſeede, which is one of the leaſt ſeedes : ſuch a patience hee had and ſetled ſpirite, to enter into the leaſt and moſt exact differences of cauſes : a fruit no doubt of the exceeding tranquillitie, and ſerenitie of his minde : which being no wayes charged or incombred, either with feares, remorſes, or ſcruples, but hauing been noted for a man of the pureſt goodneſſe without all fiction or affectation, that hath raigned or liued : made his minde continually preſent and entier : he likewiſe approached a degree neerer vnto Chriſtianitie, and became as *Agrippa* ſayd vnto S. *Paule*, *Halfe a Chriſtian* ; holding their Religion and Law in good opinion : and not only ceaſing perſecution, but giuing way to the aduancement of Chriſtians.

There ſucceeded him the firſt *Diui fratres*, the two adoptiue brethren, *Lucius Commodus Verus*, Sonne to *Elius Verus*, who delighted much in the ſofter kind of learning : and was wont to call the Poet Martiall his *Virgill* : and *Marcus Aurelius Antoninus*, whereof the later, who obſcured his colleague, and ſuruiued him long, was named the Philoſopher : who as he excelle d all the reſt in learning, ſo he excelled them likewiſe in perfection of all royall vertues ;

tues:infomuch as *Iulianus* the Emperor in his booke intituled, *Cæfares*, being as a Pafquill or Satyre, to deride all his Predeceffors, fayned that they were all inuited to a banquet of the Gods, and *Sylenus* the Iefter fate at the neather end of the table, and beftowed a fcoffe on euerie one as they came in, but when *Marcus Philofophus* came in, *Sylenus* was grauelled, and out of countenance, not knowing where to carpe at him, faue at the laft, he gaue a glaunce at his patience towards his wife. And the vertue of this Prince continued with that of his Predeceffor made the name of *Antoninus* fo facred in the world, that though it were extreamely difhonoured in *Commodus, Caracalla*, and *Hæliogabalus*, who all bare the name, yet when *Alexander Seuerus* refufed the name, becaufe he was a ftranger to the familie, the *S*enate with one acclamation fayd, *Quomodo Auguftus fic & Antoninus*. In fuch renowne and veneration, was the name of thefe two Princes in thofe dayes, that they would haue had it as a perpetuall addition in all the Emperours ftile. In this Emperours time alfo, the Church for the moft part was in peace, fo as in this fequence of fixe Princes, we doe fee the bleffed effects of Learning in foueraigntie, painted forth in the greateft Table of world.

But for a Tablet or picture of fmaller volume (not prefuming to fpeake of your Maieftie that liueth) in my iudgement the moft excellent, is that of Queene *Elizabeth*, your immediate Predeceffor in this part of *Brittaine*, a Prince, that if *Plutarch* were

now

now aliue to write lynes by parallells, would trouble him I thinke, to find for her a parallell amongst women. This Ladie was endued with learning in her sexe singuler; and grace euen amongst masculine Princes: whether we speake of Learning, of Language, or of science, moderne, or ancient: Diuinitie or Humanitie. And vnto the verie last yeare of her life, she accustomed to appoint set houres for reading, scarcely any young Student in an Vniuersitie, more dayly, or more duly. As for the gouernement, I assure my selfe, I shall not exceed, if I doe affirme, that this part of the Iland, neuer had 45. yeres of better times; and yet not through the calmnesse of the season; but through the wisedom of her regimēt. For if there be considered of the one side, the truth of Religion established; the constant peace and securitie: the good administration of Iustice, the temperate vse of the prerogatiue, not slackened, nor much strayned : the flourishing state of Learning, sortable to so excellent a Patronesse; the conuenient estate of wealth and meanes, both of Crowne and subiect : the habite of obedience, and the moderation of discontents: and there be considered on the other side, the differences of Religion, the troubles of Neighbour Countreys, the ambition of *Spaine*, and opposition of *Rome*, and then, that shee was solitary, and of her selfe : these things I say considered : as I could not haue chosen an instance so recent and so proper : so, I suppose, I could not haue chosen one more remarqueable, or eminent, to the purpose nowe

K   in

in hand; which is concerning the coniunction of learning in the Prince, with felicitie in the people.

Neither hath Learning an influence and operation onely vpon ciuile merit and morall vertue; and the Arts or temperature of peace, and peaceable gouernement; but likewife it hath no leffe power and efficacie in inablement towards martiall and militarie vertue and proweffe; as may be notably reprefented in the examples of *Alexander* the Great, and *Cæfar* the Dictator mentioned before, but now in fit place to bee refumed, of whofe vertues and Acts in warre, there needes no note or recitall, hauing beene the wonders of time in that kind. But of their affections towardes learning, and perfections in learning, it is pertinent to fay fomewhat.

Alexander was bred and taught vnder *Ariftotle* the great Philofopher; who dedicated diuers of his Bookes of Philofophie vnto him; he was attended with *Callifthenes*, and diuers other learned perfons, that followed him in Campe, throughout his Iourneyes and Conquefts: what price and eftimation hee had learning in, doth notably appeare in thefe three particulars: Firft, in the enuie he vfed to expreffe, that he bare towards *Achilles*, in this, that he had fo good a Trumpet of his prayfes as *Homers* verfes: Secondly, in the iudgement or folution he gaue touching that precious Cabinet of *Darius* which was found among his Iewels, whereof queftion was made, what thing was worthy to be put into it, and he gaue his opinion for *Homers* workes. Thirdly, in
his

his letter to *Aristotle* after hee had set forth his Bookes of Nature; wherein he expostulateth with him for publishing the secrets or misteries of Philosophie, and gaue him to vnderstand that himselfe esteemed it more to excell other men in learning & knowledge, than in power and Empire. And what vse he had of learning, doth appeare, or rather shine in all his speeches and answeres, being full of science and vse of science, and that in all varietie.

And herein againe, it may seeme a thing scholasticall, and somewhat idle to recite things that euery man knoweth; but yet, since the argument I handle leadeth mee thereunto, I am glad that men shall perceiue I am as willing to flatter (if they will so call it) an *Alexander*, or a *Cæsar*, or an *Antoninus*, that are dead many hundreth yeeres since, as any that now liueth : for it is the displaying of the glorie of Learning in Soueraigntie that I propound to my selfe, and not an humour of declayming in any mans praises. Obserue then the speech hee vsed of *Diogenes*, and see if it tend not to the true state of one of the greatest questions of morall Philosophie; whether the enioying of outward thinges, or the contemning of them be the greatest happinesse ; for when he saw *Diogenes* so perfectly contented with so little : he sayd to those that mocked at his condition: *Were I not Alexander, I would wish to be Diogenes.* But *Seneca* inuerteth it, and sayth; *Plus erat, quod hic nollet accipere, quam quod ille posset dare. There were more things which Diogenes would haue refused, tha*

K 2 *those*

thofe were which *Alexander* could haue giuen or enioyed.
Obferue again that fpeech which was vfuall with
him, *That hee felt his mortality chiefely in two thinges,*
Sleepe & Luft: & fee if it were not a fpeech extracted
out of the depth of naturall Philofophie, and liker
to haue comen out of the mouth of *Ariftotle*, or
Democritus, than from *Alexander*.

See againe that fpeech of Humanitie and poefie:
when vppon the bleeding of his wounds, he called
vnto him one of his flatterers, that was wont to a-
fcribe to him diuine honor, and faid, *Looke, this is very*
blood: this is not fuch a liquor as Homer fpeaketh of, which
ran from Venus hand, when it was pierced by Diomedes.

See likewife his readineffe in reprehenfion of Logique, in the fpeech hee vfed to *Caffander*, vppon a
complaint that was made againft his Father *Antipa-*
ter: for when *Alexander* happed to fay : *Doe you*
thinke thefe men would haue come from fo farre to
complaine, except they had iuft caufe of griefe? and
Caffander anfwered, *Yea, that was the matter, becaufe*
they thought they fhould not be difprooued; fayd *Alex-*
ander laughing: *See the fubtilties of Ariftotle, to take*
a matter both wayes, Pro & Contra, &c.

But note againe how well he could vfe the fame
Art, which hee reprehended to ferue his owne
humor, when bearing a fecret grudge to *Callifthenes*,
becaufe he was againft the new ceremonie of his a-
doration : feafting one night, where the fame *Callif-*
henes was at the table: it was mooued by fome after
fupper, for entertainement fake, that *Callifthenes* who
was

was an eloquent man, might speake of some theame or purpose at his owne choise, which *Callisthenes* did; chusing the praise of the Macedonian Nation for his discourse, & performing the same with so good maner, as the hearers were much rauished: wherupon *Alexander* nothing pleased, sayd: *It was easie to be eloquent, vpon so good a subiect*: But saith hee, *Turne your stile, and let vs heare what you can say against vs*: which *Callisthenes* presently vndertooke, and did with that stinge & life, that *Alexander* interrupted him, & sayd: *The goodnesse of the cause made him eloquent before: and dispight made him eloquent then againe.*

Consider further, for tropes of Rhetorique, that excellent vse of a Metaphor or translation, wherewith he taxed *Antipater*, who was an imperious and tyrannous Gouernor: for when one of *Antipaters* friends commended him to *Alexander* for his moderation; that he did not degenerate, as his other Lieftenants did into the Persian pride, in vse of purple; but kept the anciet habit of Macedon, of black; *True* (saith *Alexander*) *but Antipater is all purple within.* Or that other, when *Parmenio* came to him in the plaine of *Arbella*, and shewed him the innumerable multitude of his enemies specially as they appeared by the infinite number of lights; as it had beene a new firmament of starres; and thereupon aduised him to assayle them by night whereupon he answered, *That he would not steale the Victorie.*

For matter of pollicie, weigh that significant distinction so much in al ages embraced, that he made between his two friends *Epheftion* and *Craterus*, whe he
sayd,

fayd, *That the one loued Alexander, and the other loued the King* ; defcribing the principall difference of Princes beft feruants, that fome in affection loue their perfon, and other in dutie loue their crowne.

Weigh alfo that excellent taxation of an Errour ordinarie with counfellors of Princes, that they counfell their Maifters according to the modell of their owne mind and fortune, and not of their Mafters, when vpon *Darius* great offers *Parmenio* had faid : *Surely, I would accept thefe offers were I as Alexander* : fayth *Alexander* , *So would I, were I as Parmenio.*

Laftly, weigh that quicke and acute reply, which he made when he gaue fo large gifts to his friends, & feruants, and was asked what he did referue for himfelfe, and he anfwered, *Hope* : Weigh I fay, whether he had not caft vp his account aright, becaufe *Hope* muft bee the portion of all that refolue vppon great enterprifes. For this was *Cæfars* portion, when he went firft into *Gaule*, his eftate being then vtterly ouerthrowne with Largeffes : And this was likewife the portion of that noble Prince, howfoeuer tranfported with ambition, *Henry* Duke of *Guife*, of whom it was vfually fayd : that he was the greateft Vfurer in *Fraunce*, becaufe he had turned all his eftate into obligations.

To conclude therefore, as certaine *Critiques* are vfed to fay hyperbolically: *That if all Sciences were loft, they might bee found in Virgill*: So certainely this may be fayd truely ; there are the prints, and footeﬅeps

steps of learning in those fewe speeches, which are reported of this Prince. The admiration of whom, when I consider him, not as *Alexander* the Great, but as *Aristotles* Scholler, hath carryed me too farre.

As for *Iulius Cæsar*, the excellencie of his learning, needeth not to be argued from his education, or his companie, or his speeches : but in a further degree doth declare it selfe in his writinges and workes, whereof some are extant, and permanent, and some vnfortunately perished : For, first we see there is left vnto vs that excellent Historie of his owne warres, which he entituled onely a Commentarie, wherin all succeeding times haue admired the solide weight of matter, and the reall passages, and liuely Images of actions, and persons expressed in the greatest proprietie of words, and perspicuitie of Narration that euer was : which that it was not the effect of a naturall guift, but of learning and precept, is well witnessed by that worke of his, entituled *De Analogia*, being a grammaticall Philosophie, wherein hee did labour to make this same *Vox ad placitum*, to become *Vox ad licitum* : and to reduce custome of speech, to congruitie of speeeh, and tooke as it were the pictures of wordes, from the life of reason.

So wee receiue from him as a Monument, both of his power and learning, the then reformed computation of the yeare, well expressing, that he tooke it to be as great a glorie to himselfe, to obserue and know the law of the heauens, as to giue law to men vpon the earth.

So likewife in that booke of his *Anticato*, it may eafily appeare that he did afpire as well to victorie of of wit, as victory of warre : vndertaking therein a conflict againft the greateft Champion with the pen that then liued, *Cicero* the Orator.

So againe in his Booke of *Apothegmes*, which he collected, we fee that he efteemed it more honour to make himfelfe, but a paire of Tables, to take the wife and pithy words of others, than to haue euery word of his owne to be made an Apothegme, or an Oracle ; as vaine Princes, by cuftome of flatterie, pretend to doe. And yet if I fhould enumerate diuers of his fpeeches ; as I did thofe of *Alexander*, they are truely fuch as *Salomon* noteth, when hee fayth; *Verba fapientum tanquam aculei, & tanquam claui in altum defixi*, whereof I will only recite three, not fo delectable for elegancie, but admirable for vigor and efficacie.

As firft, it is reafon hee bee thought a Mafter of words, that could with one word appeafe a mutinie in his Armie; which was thus. The Romanes when their Generals did fpeake to their Armie, did vfe the word *Milites;* but when the Magiftrates fpake to the people, they did vfe the word, *Quirites :* The Souldiers were in tumult, and feditioufly prayed to bee cafsiered : not that they fo meant, but by expoftulation thereof, to drawe *Cæfar* to other Conditions ; wherein hee being refolute, not to giue way, after fome filence, hee beganne his fpeech, *Ego Quirites,* which did admit them alreadie cafsiered ; where-

with

with they were ſo ſurpriſed, croſſed, and confuſed, as they would not ſuffer him to goe on in his ſpeech, but relinquiſhed their demaunds, and made it their ſuit, to be againe called by the name of *Milites*.

The ſecond ſpeech was thus : *Cæſar* did extreamly affect the name of King; and ſome were ſet on as he paſſed by, in popular acclamation to ſalute him king; whereupon finding the crie weake and poore; he put it off thus, in a kind of Ieſt, as if they had miſtaken his ſurname; *Non Rex ſum, ſed Cæſar*, a ſpeech, that if it be ſearched, the life and fulneſſe of it, can ſcarce be expreſſed : For firſt it was a refuſall of the name, but yet not ſerious : againe it did ſignifie an infinite confidence and magnanimitie, as if he preſumed *Cæſar* was the greater title ; as by his worthineſſe, it is come to paſſe till this day: but chiefely, it was a ſpeech of great allurement toward his owne purpoſe: as if the State did ſtriue with him, but for a name; whereof meane families were veſted : for *Rex* was a ſurname with the *Romanes*, aſwell as *King* is with vs.

The laſt ſpeech, which I will mention, was vſed to *Metellus :* when *Cæſar*, after warre declared, did poſſeſſe himſelfe of the Citie of *Rome*, at which time entring into the inner treaſurie, to take the the monney there accumulate, *Metellus* being Tribune forbad him : whereto *Cæſar* ſayd, *That if hee did not deſiſt, hee would laye him dead in the place :* And preſently taking himſelfe vp, hee added : *Young man it is harder for me to ſpeake it, than to doe it ; Adolefcens, durius eſt mihi, hoc dicere, quàm facere.* A ſpeech compounded of the greateſt terrour, and greateſt clemencie, that could proceede out of the mouth of man.

But to returne and conclude with him, it is euident himſelfe knewe well his owne perfection in learning, and tooke it vpon him ; as appeared, when vpon occaſion, that ſome ſpake, what a ſtrange reſolution it was in *Lucius Sylla,* to reſigne his Dictature; he ſcoffing at him, to his owne aduantage, anſwered; *That Sylla could not skill of Letters, and therefore knew not how to Dictate.*

And here it were fit to leaue this point, touching the concurrence of militarie vertue and learning; (for what example ſhould come with any grace, after thoſe two, of *Alexander* and *Cæſar*) were it not in regard of the rareneſſe of circumſtance, that I finde in one other particular; as that which did ſo ſodenly paſſe, from extreame ſcorne, to extreame wonder : and it is of *Xenophon* the Philoſopher, who went from *Socrates* Schoole into *Aſia,* in the expedition of *Cyrus* the younger, againſt King *Artaxerxes :* This *Xenophon* at that time, was verie yong, and neuer had ſeene the Warres before : neither had any commaund in the Armie, but onely followed the Warre, as a voluntarie, for the loue and conuerſation of *Proxenus* his friend : hee was preſent when *Falinus* came in Meſſage from the great King, to the Grecians ; after that *Cyrus* was ſlaine in the field; and they a handfull of men left to themſelues

in

in the middeſt of the Kings Territories, cut off from their Country by many nauigable Riuers, and many hundred miles: The Meſſage imported, that they ſhould deliuer vp their Armes, and ſubmit themſelues to the Kings mercy: To which Meſſage before anſwere was made, diuers of the Army cõferred familiarly with *Falinus;* and amongſt the reſt *Xenophon* happened to ſay: *Why Falinus, we haue now but theſe two thinges left; our Armes, and our Vertue: and if we yeeld vp our Armes, how ſhall we make vſe of our Vertue?* Whereto *Falinus* ſmiling on him, ſayd; *If I be not deceiued, young Gentleman, you are an Athenian; and I beleeue, you ſtudie Philoſophie, and it is pretty that you ſay; but you are much abuſed, if you thinke your vertue can withſtand the Kings power:* Here was the ſcorne; the wonder followed; which was, that this young Scholler, or Philoſopher, after all the Captaines were murthered in parlye by treaſon, conducted thoſe ten Thouſand foote, through the heart of all the Kinges high Countreys from *Babilon* to *Grecia* in ſafetie, in deſpight of all the Kings forces, to the aſtoniſhment of the world, and the encouragement of the Grecians in times ſucceeding, to make inuaſion vpon the Kings of *Perſia;* as was after purpoſed by *Iaſon* the Theſſalian; attempted by *Ageſi'aus* the Spartan, and atchieued by *Alexander* the Macedonian, all, vpon the ground of the Act of that young Scholler.

To proceede now from imperiall and militarie vertue, to morall and priuate vertue; firſt, it is an aſſured

ſured

fured truth, which is contained in the verfes;

Scilicet ingenuas didiciffe fideliter artes,
Emollit mores nec finit effe feros.

It taketh away the wildneffe and barbarifme and
fierceneffe of mens minds: but indeed the accent had
need be vpon, *fideliter.* For a little fuperficiall learning doth rather worke a contrary effect. It taketh
away all leuitie, temeritie, and infolencie, by copi-
ous fuggeftion of all doubts and difficulties, and
acquainting the minde to ballance reafons on both
fides, and to turne backe the firft offers and conceits of the minde, and to accept of nothing but
examined and tryed. It taketh away vaine admiration of any thing, which is the roote of all
weakeneffe. For all things are admired, either becaufe they are new, or becaufe they are great. For
nouelty, no man that wadeth in learning or contemplation throughly, but will find that printed in his
heart, *Nil noui fuper terram* : Neither can any man
maruaile at the play of Puppets, that goeth behinde
the curtaine, and aduifeth well of the Motion. And
for magnitude, as *Alexander* the Great, after
that hee was vfed to great Armies, and the great
Conquefts of the fpatious Prouinces in *Afia,*
when hee receiued Letters out of *Greece,* of fome
fights and feruices there, which were commonly
for a paffage, or a Fort, or fome walled *Towne*
at the moft, he fayd ; *It feemed to him, that he was*
aduertifed of the battailes of the Frogs, and the Mife,
that the ould tales went of. So certainely, if a man
meditate

meditate much vppon the vniuerſall frame of nature, the earth with men vppon it (the diuineſſe of
ſoules except) will not ſeeme much other, than
an Ant hill, whereas ſome Ants carrie corne, and
ſome carrie their young: and ſome goe emptie,
and all too and fro, a little heape of duſt. It taketh
away, or mitigateth feare of death, or aduerſe fortune: which is one of the greateſt impediments
of vertue, and imperfections of manners. For
if a mans minde, be deepely ſeaſoned with the conſideration of the mortalitie and corruptible
nature of thinges, hee will eaſily concurre with
Epictetus, who went foorth one day, and ſawe a
woman weeping for her Pitcher of earth, that
was broken ; and went foorth the next day,
and ſawe a woman weepinge for her Sonne
that was deade, and thereuppon ſayde : *He-*
ri, vidi fragilem frangi, hodie vidi mortalem mo-
ri. And therefore *Virgill* did excellently, and
profoundlye couple the knowledge of cauſes,
and the Conqueſt of all feares, together, as
Concomitantia.

 Fœlix qui potuit rerum cognoſcere cauſas,
 Quique metus omnes, & inexorabile fatum
 Subiecit pedibus, ſtrepitumque Acherontis auari.

It were too long to goe ouer the particnlar remedies, whieh learning doth miniſter, to all the diſeaſes of the minde, ſometimes purging the ill humours, ſometimes opening the obſtructions, ſometimes helping digeſtion, ſometimes encreaſing
   L 3   appetite,

appetite, fomtimes healing the wounds and exulcerations thereof, and the like; and therefore I will
conclude with that which hath *rationem totius;*
which is, that it difpofeth the conftitution of the
minde, not to be fixed or fetled in the defects thereof; but ftill to be capable, and fufceptible of growth
and reformation. For the vnlearned man knowes
not, what it is to defcend into himfelfe, or to cal himfelfe to account, nor the pleafure of that *Suauiſsima
vita, indies fentire ſe fieri meliorem*: The good parts
hee hath, hee will learne to ſhew to the full, and vſe
them dexteroufly, but not much to encreafe them:
The faults he hath, he will learne how to hide and
colour them, but not much to amend them; like an
ill Mower, that mowes on ftill, and neuer whets his
Syth: whereas, with the learned man, it fares otherwife, that he doth euer intermix the correction and
amendment of his minde, with the vfe and employment thereof: Nay further in generall and in fum :
certain it is, that *Veritas,* and *Bonitas* differ, but as the
Seale and the Print: for Truth prints Goodneffe,
and they be the cloudes of Error, which defcend in
the ſtormes of paſsions and perturbations.

From morall vertue, let vs paffe on to matter of
power and commandement, and confider whether
in right reafon, there be any comparable with that,
wherewith knowledge inuefteth and crowneth
mans nature. We fee the dignitie of the commandement, is according to the dignitie of the commaunded: to haue commaundement ouer beafts, as
Heard-

Heard-men haue, is a thing contemptible: to haue commandement ouer children, as Schoole-Maſters haue, is a matter of ſmall honor : to haue commandement ouer Gally-ſlaues, is a diſparagement, rather than an honour. Neither is the commaundement of *T*yrants, much better ouer people, which haue put off the Generoſitie of their mindes : And therefore it was euer holden, that honors in free Monarchies and Common wealths, had a ſweetneſſe more, than in Tyrannies, becauſe the commandement extendeth more ouer the wils of men, and not only ouer their deeds and ſeruices. And therefore when *Virgill* putteth himſelfe forth to attribute to *Auguſtus Cæſar* the beſt of humane honours, hee doth it in theſe wordes :

Victorque volentes
Per populos, dat iura, viamque affectat Olympo:

But yet the commandement of knowledge, is yet higher, than the commandement ouer the will: for it is a commaundement ouer the reaſon, beleefe, and vnderſtanding of man, which is the higheſt part of the minde, and giueth law to the will it ſelfe. For there is no power on earth, which ſetteth vp a throne or chaire of Eſtate in the ſpirits, and ſoules of men, and in their cogitations, imaginations, opinions, and beleefes : but knowledge and learning. And therefore wee ſee the deteſtable and extreame pleaſure, that Arch heretiques, and falſe Prophets, and Impoſtors are tranſported with, when they once finde in themſelues, that they haue a ſuperioritie in the faith

and

and confcience of men; fo great, as if they haue once tafted of it, it is feldome feene, that any torture or perfecution can make them relinquifh or abandone it. But as this is that which the Author of the Reuelation, calleth the depth or profoundneffe of Sathan: fo by argument of contraries, the iuft and lawfull foueraignetie ouer mens vnderftanding, by face of truth rightly interpreted, is that which approacheth neereft to the fimilitude of the diuine rule.

As for fortune and aduancement, the beneficence of learning, is not fo côfined to giue fortune only to ftates and Common-wealthes : as it doth not likewife giue fortune to particular perfons. For it was well noted long agoe, that *Homer* hath giuen more men their liuings, than either *Sylla*, or *Cæfar*, or *Auguftus* euer did, notwithftanding their great largeffes, and donatiues, and diftributions of Lands to fo many legions. And no doubt, it is hard to fay, whether armes or learning haue aduanced greater numbers. And in cafe of foueraigntie, wee fee, that if armes or defcent haue carried away the *K*ingdome: yet learning hath carryed the Prieft hood, which euer hath been in fome competicion with Empire.

Againe, for the pleafure and delight of knowledge and learning, it farre furpaffeth all other in nature: for fhall the pleafures of the affections fo exceede the fences, as much as the obtayning of defire or victorie, exceedeth a fong, or a dinner ? and muft not of confequence, the pleafures of the intellect

lect or vnderſtanding exceede the pleaſures of the affections? we ſee in all other pleaſures, there is ſacietie; and after they be vſed, their verdour departeth; which ſheweth well, they be but deceits of pleaſure, and not pleaſures; and that it was the noueltie which pleaſed, and not the qualitie. And therfore we ſee, that voluptuous men turne Friers; and ambitious Princes turne melancholy. But of knowledge there is no ſacietie, but ſatisfaction and appetite, are perpetually interchangeable; and therefore appeareth to be good in it ſelfe ſimply, without fallacie or accident. Neither is that pleaſure of ſmall efficacie, and contentment to the minde of man, which the Poet *Lucretius* deſcribeth elegantly,

Suaue mari magno, turbantibus æquora ventis: &c.

It is a view of delight (ſayth he) *to ſtand or walke vppon the ſhoare ſide, and to ſee a Shippe toſſed with tempeſt vpon the ſea; or to bee in a fortified Tower, and to ſee two Battailes ioyne vppon a plaine. But it is a pleaſure incomparable for the minde of man to bee ſetled, landed, and fortified in the certaintie of truth; and from thence to deſcrie and behould the errours, perturbations, labours, and wanderings vp and downe of other men.*

Laſtly, leauing the vulgar arguments, that by learning, man excelleth man in that, wherein man excelleth beaſts; that by learning man aſcendeth to the heauens and their motions; where in bodie he cannot come; and the like; let vs conclude with the dignitie and excellency of knowledge and learning, in that whereunto mans nature doth moſt aſpire;

M which

which is immortalitie or continuance ; for to this
tendeth generation, and rayfing of houfes and families ; to this buildings, foundations, and monuments,
to this tendeth the defire of memorie, fame, and ce-
lebration; and in effe&, the ftrength of all other humane defires ; wee fee then howe farre the monuments of wit and learning, are more durable, than
the monuments of power, or of the hands. For haue
not the verfes of *Homer* continued 25. hundred
yeares, or more, without the loffe of a fillable, or
letter: during which time, infinite Pallaces, Temples, Caftles, Cities haue been decayed, and demo-
lifhed ? It is not pofsible to haue the true pictures or
ftatuaes of *Cyrus*, *Alexander*, *Cæfar*, no nor of the
Kings, or great perfonages of much later yeares;
for the originals cannot laft; and the copies cannot
but leefe of the life and truth. But the Images of
mens wits and knowledges remaine in Bookes, exempted from the wrong of time, and capable of perpetuall renouation : Neither are they fitly to be called Images, becaufe they generate ftill, and caft their
feedes in the mindes of others, prouoking and caufing infinit actions and opinions, in fucceeding ages.
So that if the inuention of the Shippe was thought
fo noble, which carryeth riches, and commodities
from place to place, and confociateth the moft remote regions in participation of their fruits : how
much more are letters to bee magnified, which as
Shippes, paffe through the vaft Seas of time, and
make ages fo diftant, to participate of the wifedome,

illumi-

illuminations, and inuentions the one of the other?
Nay further wee see, some of the Philosophers
which were least diuine, and most immersed in the
sences, and denyed generally the immortality of the
soule; yet came to this point, that whatsoeuer motions the spirite of man could act, and perfourme
without the Organs of the bodie, they thought
might remaine after death; which were only those
of the vnderstanding, and not of the affection; so
immortall and incorruptible a thing did knowledge
seeme vnto them to be. But we that know by diuine
reuelation, that not onely the vnderstanding, but
the affections purified, not onely the spirite,
but the bodie changed shall be aduanced to immortalitie, doe disclaime in these rudiments of the sences. But it must be remēbred, both in this last point,
and so it may likewise be needfull in other places,
that in probation of the dignitie of knowledge, or
learning I did in the beginning separate diuine testimonie, from humane; which methode, I haue pursued, and so handled them both apart.

Neuerthelesse, I doe not pretend, and I know it
will be impossible for me by any Pleading of mine,
to reuerse the iudgement, either of *Æsops* Cocke,
that preferred the Barly-corne, before the Gemme;
or of *Mydas,* that being chosen Iudge, betweene
Apollo President of the Muses, and *Pan* God of the
Flockes, iudged for Plentie: or of *Paris,* that iudged for Beautie, and loue against Wisedome and
Power: nor of *Agrippina, Occidat matrem, modo im-*

peret:

peret: that preferred Empire with condition neuer
so deteſtable; or of *Vlyſſes, Qui vetulam prætulit im-
mortalitati*, being a figure of thoſe which preferre
Cuſtome and Habite before all excellencie; or of a
number of the like popular Iudgements. For theſe
thinges continue, as they haue beene: but ſo
will that alſo continue, whereupon learning hath euer relyed, and which
fayleth not: *Iuſtificata eſt ſa-
pientia à filij s ſuis.*

THE SECOND

Booke of FRANCIS BACON; of
the proficience or aduancement
of Learning , Diuine and
Humane.

To the King.

I T might seeme to haue more conuenience , though it come often otherwise to passe, (Excellent King)that those which are fruit-full in their generati-ons, & haue in them-selues the foresight of Immortalitie,in their descendents,should likewise be more carefull of the good estate of future times; vnto which they know they must transmitte and commend ouer their dea-rest pledges. Queene *Elizabeth* was a soiourner in the world in respect of her vnmaried life : and was a blessing to her owne times; & yet so as the impres-sion of her good gouernement, besides her happie
A a   memorie,

memorie, is not without fome effect, which doth
furuiue her. But to your Maieftie, whom God hath
alreadie bleffed with fo much Royall iffue, worthie
to continue and reprefent you for euer : and whofe
ycuthfull and fruitfull bedde doth yet promife manie the like renouations: It is proper and agreeable
to be conuerfant, not only in the tranfitory parts of
good gouernment : but in thofe acts alfo, which are
in their nature permanent & perpetuall. Amongft
the which (if affection do not tranfport mee,) there
is not any more worthie, then the further endowe-
ment of the world with found and fruitfull knowledge : For why fhould a fewe receiued Authors
ftand vp like *Hercules Columnes*, beyond which, there
fhould be no fayling, or difcouering, fince wee haue
fo bright and benigne a ftarre, as your Ma: to conduct and profper vs ? To returne therefore where
wee left, it remaineth to confider of what kind thofe
Acts are which haue bene vndertaken, & performed
by Kings and others, for the increafe and aduancement of learning, wherein I purpofe to fpeake ac-
tiuely without digreffing or dylating.

Let this ground therfore be layd, that all workes
are ouercómen by amplitude of reward, by foundneffe of direction, and by the coniunction of labors.
The firft multiplyeth endeuour, the fecond preuen-
teth error, and the third fupplieth the frailty of man.
But the principal of thefe is direction: For *Claudus in
via, anteuertit curforem extra viam* : And *Salomon*
excellently fetteth it downe ; *If the Iron be not fharpe,*

it

it requireth more strength: But wisedome is that which preuaileth: signifying that the Inuention or election of the Meane, is more effectuall then anie inforcement or accumulation of endeuours. This I am induced to speake; for that(not derogating from the noble intention of any that haue beene deseruers towards, the State of learning)I do obserue neuerthelesse, that their workes and Acts are rather matters of Magnificence and Memorie, then of progression and proficience, and tende rather to augment the masse of Learning in the multitude of learned men, then to rectifie or raise the Sciences themselues.

The Works or Acts of merit towards learning are conuersant about three obiects, the Places of learning; the Bookes of learning; and the Persons of the learned. For as water,whether it be the dewe of heauen,or the springs of the earth, doth scatter and leese it selfe in the ground, except it be collected into some Receptacle,where it may by vnion,comfort and sustaine it selfe: And for that cause the Industry of Man hath made & framed Spring heads, Conduits, Cesternes,and Pooles,which men haue accustomed likewise to beautifie and adorne with accomplishments of Magnificence and State,as wel as of vse and necessitie: So this excellent liquor of knowledge, whether it descend from diuine inspiration,or spring from humane sense, would soone perishe and vanishe to oblyuion, if it were not preserued in Bookes,Traditions,Conferences,and

Places appoynted, as Vniuerfities, Colledges, and Schooles, for the receipt & comforting of the fame.

The works which concerne the Seates and Places of learning, are foure; Foundations, and Buyldings, Endowments with Reuenewes, Endowmēts with Franchizes and Priuiledges, Inftitutions and Ordinances for gouernment, all tending to quietnefle and priuatenefle of life, and difcharge of cares and troubles, much like the Stations, which *Virgil* prefcribeth for the hyuing of Bees.

Principio fedes Apibus, ftatioq; petenda :
    Quo neq; fit ventis aditus, &c.

The workes touching Bookes are two: Firft Libraries, which are as the Shrynes, where all the Reliques of the ancient Saints, full of true vertue, and that without delufion or impofture, are preferued, and repofed ; Secondly, Newe Editions of Authors, with more correct impreffions, more faithfull tranflations, more profitable gloffes, more diligent annotations, and the like.

The workes pertaining to the perfons of learned men (befides the aduancement and countenancing of them in generall) are two : The reward and defignation of Readers in Sciences already extant and inuented : and the reward and defignation of Writers and Enquirers, concerning any partes of Learning, not fufficiently laboured and profecuted.

Thefe are fummarilie the workes and actes, wherein the merites of manie excellent Princes, and other worthie Perfonages haue beene conuerfant.

fant. As for any particular commemorations, I call to minde what *Cicero* faide, when hee gaue generall thanks. *Difficile non aliquem;ingratum quenquam præterire* : Let vs rather according to the Scriptures, looke vnto that parte of the Race , which is before vs ; then looke backe to that which is alreadie attained.

Firft therfore amongft fo many great Foundations of Colledges in *Europe,* I finde ftrange that they are all dedicated to Profeffions,and none left free to Artes and Sciences at large. For if men iudge that learning fhould bee referred to action , they iudge well : but in this they fall into the Error defcribed in the ancient Fable; in which the other parts of the body did fuppofe the ftomache had beene ydle, becaufe it neyther performed the office of Motion, as the lymmes doe, nor of Sence, as the head doth : But yet notwithftanding it is the Stomache that digefteth and diftributeth to all the reft: So if any man thinke Philofophie and Vniuerfalitie to be idle Studies; hee doth not confider that all Profeffions are from thence ferued, and fupplyed. And this I take to bee a great caufe that hath hindered the progreffion of learning,becaufe thefe Fundamental knowledges haue bene ftudied but in paffage. For if you will haue a tree beare more fruite then it hath vfed to do; it is not any thing you can do to the boughes, but it is the ftyrring of the earth, and putting newe moulde about the rootes , that muft woike it. Neyther is it to bee forgotten, that this dedicating of Foundations and Dotations to profeſſory Learning, hath not onely had a Maligne aſpect, and influence vpon the growth of Scyences, but hath alſo beene preiudiciall to States and gouernments. For hence it proceedeth that Princes find a ſolitude, in regard of able men to ſerue them in cauſes of eſtate, becauſe there is no education collegiate, which is free; wher ſuch as were ſo diſpoſed, mought giue themſelues to Hiſtories, moderne languages, bookes of pollicie and ciuile diſcourſe, and other the like inablements vnto ſeruice of eſtate.

And becauſe founders of Colledges doe plant, and founders of Lectures doe water: it followeth wel in order to ſpeake of the defect, which is in Publique Lectures: Namely, in the ſmalneſſe and meaneſſe of the ſalary or reward which in moſt places is aſſigned vnto them: whether they be Lectures of Arts, or of Profeſſions. For it is neceſſary to the progreſſion of Scyences, that Readers be of the moſt able and ſufficient men; as thoſe which are ordained for generating, and propagating of Scyences, and not for tranſitorie vſe. This cannot be, except their condition, & endowmēt be ſuch, as may cōtent the ableſt man, to appropriate his whole labour, and continue his whole age in that function and attendance, and therefore muſt haue a proportion anſwerable to that mediocritie or competencie of aduancement, which may be expected from a Profeſſion, or the Practize of a Profeſſion: So as, if you wil haue Scyences flouriſh, you muſt obſerue *Dauids* military

military lawe, which was, *That those which staied with
the Carriage, should haue equall part with those which
were in the Action* : else will the carriages be ill attended : So Readers in Scyences are indeede the Gar-
dyans of the stores and prouisions of Scyences,
whence men in actiue courses are furnished, and
therefore ought to haue equall entertainment with
them; otherwise if the fathers in Scyences be of the
weakest sort, or be ill maintained.

Et Patrum invalidi referent ieiunia nati.

Another defect I note, wherin I shall neede some
Alchimist to helpe me, who call vpon men to sell
their Bookes, and to build Fornaces, quitting and
forsaking *Minerva,* and the *Muses,* as barreyne vir-
gines, and relying vpon *Vulcan* . But certaine it is,
that vnto the deepe, fruitefull, and operatiue studie
of many Scyences, specially Naturall Phylosophy
and Physicke, Bookes be not onely the Instrumen-
tals; wherein also the beneficence of men hath not
beene altogether wanting : for we see, Spheares,
Globes, Astrolabes, Mappes, and the like, haue bene
prouided, as appurtenances to Astronomy & Cos-
mography, as well as bookes: We see likewise, that
some places instituted for Physicke, haue annexed
the commoditie of Gardeins for Simples of all
sorts, and do likewise command the vse of dead Bodies for Anatomyes. But these doe respect but a few
things. In generall, there will hardly be any Mayne
proficience in the disclosing of nature, except there
be some allowance for expences about experiméts;
whether

whether they be experiments appertaining to *Vul-canus* or *Dedalus*, Furnace or Engyne, or any other kind; And therefore as Secretaries, and Spyalls of Princes and States bring in Bills for Intelligence; so you must allowe the Spyalls and Intelligencers of Nature, to bring in their Billes, or else you shall be ill aduertised.

And if *Alexander* made such a liberall assignation to *Aristotle* of treasure for the allowance of Hunters, Fowlers, Fishers and the like, that he mought compile an Historie of Nature, much better do they deserue it that trauailes in Arts of nature.

Another defect which I note, is an intermission or neglect in those which are Gouernours in Vniuersities, of Consultation, & in Princes or superior persons, of Visitation : To enter into account and consideration, whether the Readings, exercises, and other customes appertayning vnto learning, aunciently begunne, and since continued, be well instituted or no, and thereupon to ground an amendement, or reformation in that which shall be found inconuenient . For it is one of your Maiesties owne most wise and Princely Maximes, *that in all vsages, and Presidents, the Times be considered wherein they first beganne, which if they were weake, or ignorant, it derogateth from the Authoritie of the Vsage, and leaueth it for suspect*. And therefore in as much, as most of the vsages, and orders of the Vniuersities were deriued frō more obscure times, it is the more requisite, they be reexamined. In this kind I will giue an
instance

inftance or two for exáple fake, of things that are the moft obvious & familiar: The one is a matter, which though it bee ancient and generall, yet I hold to be an errour, which is, that Schollers in Vniuerfities come too foone, & too vnripe to Logicke & Rhetoricke; Arts fitter for Graduates then children, and Nouices : For thefe two rightly taken, are the graueft of Sciences, beeing the Arts of Arts, the one for Iudgement, the other for Ornament : And they be the Rules & Directions, how to fet forth & difpofe matter : & therfore for mindes emptie & vnfraught with matter, & which haue not gathered that which *Cicero* calleth *Sylua* and *Supellex*, ftuffe and varietie to beginne with thofe Artes (as if one fhould learne to weigh, or to meafure, or to painte the Winde) doth worke but this effect : that the wifedome of thofe Arts, which is great, & vniuerfal, is almoft made contemptible, & is degenerate into childifh Sophyftrie, & ridiculous affectation. And further, the vntimely learning of them hath drawen on by confequence, the fuperficiall and vnprofitable teaching & writing of them, as fitteth indeed to the capacity of childré : Another, is a lacke I finde in the exercifes vfed in the Vniuerfities, which do make to great a diuorce betweene Inuention & Memory: for their fpeeches are either premeditate in *Verbis conceptis*, where nothing is left to Inuention, or meerly *Extemporall*, where little is left to Memory: wheras in life & action, there is leaft vfe of either of thefe, but ratherof intermixtures of premeditation, & Inuention: Notes & Memorie.

B b   So

So as the exercife fitteth not the practize, nor the I-
mage, the life; and it is euer a true rule in exercifes,
that they bee framed as nere as may be to the life of
practife, for otherwife they do peruert the Motions,
and faculties of the Minde, and not prepare them.
The truth whereof is not obfcure, when Schollars
come to the practifes of profeffions, or other actions
of ciuill life, which when they fet into, this want is
foone found by themfelues, and fooner by others.
But this part touching the amendment of the Inftitutions and orders of Vniuerfities. I will conclude
with the claufe of *Cæfars* letter to *Oppius* and *Balbus*,
*Hoc quemadmodum fieri poffit, nonnulla mihi in mentem
veniunt, & multa reperiri poffunt: de ijs rebus rogo vos,
vt cogitationem fufcipiatis.*

Another defect which I note, afcendeth a little
higher then the precedent. For as the proficience of
learning confifteth much in the orders and inftitutions of Vniuerfities, in the fame States & kingdoms:
So it would bee yet more aduanced, if there were
more Intelligéce Mutual betweene the Vniuerfities
of *Europe*, then now there is. We fee, there be many
Orders and Foundatiós, which though they be deuided vnder feuerall foueraignties, & territories, yet
they take themfelues to haue a kind of contract, fraternitie, & correfpondence, one with the other, infomuch as they haue Prouincials and Generals.
And furely as Nature createth Brotherhood in Families, & Arts Mechanicall contract Brotherhoods
in communalties, and the Anoyntment of God fu-
                                                perinduceth

perinduceth a Brotherhood in Kings & Biſhops:So in like manner there cannot but bee a fraternitie in learning and illumination, relating to that P aterni-tie,which is attributed to God,who is called the Fa-ther of illuminations or lights.

The laſt defeſt which *I* wil note,is,that there hath not been, or very rarely been,any Publique Deſig-nation of Writers or Enquirers , concerning ſuch parts of knowledge,as may appeare not to haue bin alreadie ſufficiently laboured or vndertaken , vnto which point it is an Inducemēt ; to enter into a view and examination , what parts of learning haue bin proſecuted, and what omitted ; For the opinion of plentie is amongſt the cauſes of want ; and the great quantitie of Bookes maketh a ſhewe rather of ſuper-fluitie then lacke , which ſurcharge neuertheleſſe is not to be remedied by making no more bookes,but by making more good books,which as the Serpēt of *Moſes*,mought deuour the Serpēts of the Inchātors.

The remouing of all the defeſts formerly enu-merate,except the laſt, and of the aſtiue part alſo of the laſt(which is the deſignation of Writers) are *O-pera Baſilica* ; towards which the endeuors of a pri-uate man may be, but as an Image in a croſſe way; that may point at the way , but cannot goe it. But the inducing part of the latter (which is the ſuruay of Learning) , may bee ſet forwarde by priuate trauaile ; Wherefore I will now attempt to make a generall and faithfull perambulation of lear-ning,with an inquiry what parts therof lye freſh and

waſt,

waſt, and not improued & conuerted by the Induſtrie of man; to the end that ſuch a plotte made and recorded to memorie, may both miniſter light to a-nie publique deſignation : and alſo ſerue to excite voluntary endeuours; wherin neuertheleſſe my purpoſe is at this time, to note onely omiſſions and deficiences; and not to make any redargution of Errors, or incomplete proſecutions: For it is one thing to ſet forth what ground lyeth vnmanured ; and another thing to correct ill husbandry in that which is manured.

In the handling & vndertaking of which worke, I am not ignorant, what it is, that I doe now mooue and attempt, nor inſenſible of mine own weakenes, to ſuſteine my purpoſe : But my hope is, that if my extreame loue to learning carrie me too farre, I may obtaine the excuſe of affection; for that *It is not granted to man to loue, and to bee wiſe.*   But I know well I can vſe no other libertie of Iudgement, then I muſt leaue to others, & I for my part ſhall be indifferentlie glad eyther to performe my ſelfe, or accept from another, that dutie of humanitie : *Nam qui erranti comiter monſtrat viam: &c.*   I doe foreſee likewiſe, that of thoſe things, which I ſhall enter & Regiſter, as Deficiences and Omiſſions: Many will conceiue and cenſure, that ſome of them are alreadie done & extant : others to bee but curioſities, and things of no great vſe : and others to bee of too great difficultie, and almoſt impoſſibilitie to bee compaſſed and effected : But for the twoo firſt , I referre my ſelfe to
the

the particulars. For the laft, touching impofsibilitie, I take it, thofe things are to bee held pofsible, which may be done by fome perfon, though not by euerie one : and which may be done by many, though not by any one : and which may be done in fuccefsion of ages, though not within the houre-glaffe of one mans life : and which may be done by publique defignation, though not by priuate endeuour. But notwithftãding, if any Man will take to himfelfe rather that of Salomon, *Dicit piger, Leo eft in via*, then that of Virgil, *Poffunt, quia poffe videntur* : I fhall be content that my labours bee efteemed, but as the better forte of wifhes : for as it asketh fome knowledge to demaund a queftion, not impertinent ; fo it requireth fome fenfe, to make a wifh not abfurd.

THE PARTS of humane learning haue reference to the three partes of Mans vnderftanding, which is the feate of Learning: HISTORY to his MEMORY, POESIE to his IMAGINATION, and PHILOSOPHIE to his REASON: Diuine learning receiueth the fame diftribution, for the Spirit of Man is the fame : though the Reuelation of Oracle and Senfe be diuerfe : So as Theologie confifteth alfo of HISTORIE of the Church ; of PARABLES, which is Diuine *Poefie*: and of holie DOCTRINE or *Precept*. For as for that part, which feemeth fupernumerarie, which is *Prophecie* : it is but Diuine Hiftorie : which hath that prerogatiue ouer humane, as the Narration may bee before the fact, afwell as after.

Of the aduancement of learning

HISTORY is NATVRALL, CIVILE, ECCLESIASTICALL & LITERARY, wherof the three first I allow as extant, the fourth I note as deficient. For no man hath propounded to himselfe the generall state of learning to bee described and rcpresented from age to age, as many haue done the works of Nature, & the State ciuile and Ecclesiastical; without which the History of the world seemeth to me, to be as the *Statua* of *Polyphemus* with his eye out, that part being wanting, which doth most shew the spirit, and life of the person : And yet I am not ignorant that in diuers particular sciences, as of the Iurisconsults, the Mathematicians, the Rhetoricians the Philosophers, there are set down some smal memorials of the Schooles, Authors, and Bookes: and so likewise some barren relations touching the Inuentió of Arts, or vsages. But a iust story of learning, containing the Antiquities & Originalls of Knowledges, & their Sects, their Inuentions, their Traditions; their diuerse Administrations, and Managings; their Flourishings , their Oppositions, Decayes, Depressions, Obliuions, Remoues; with the causes, and occasions of them, and all other euents concerning learning, throughout the ages of the world; I may truly affirme to be wanting. The vse and end of which worke, I doe not so much designe for curiositie, or satisfaction of those that are the louers of learning ; but chiefely for a more serious, & graue purpose, which is this in fewe wordes, that it will
make

make learned men wife, in the vfe and adminiſtration of learning. For it is not Saint *Auguſtines*
nor Saint *Ambroſe* workes that will make fo wiſe
a Diuine, as Eccleſiaſticall Hiſtorie, throughly read
and obferued : and the fame reaſon is of Learning.

HISTORY of NATVRE is of three forts: of
NATVRE in COVRSE; of NATVRE ER-
RING, or VARYING; and of NATVRE AL-
TERED or wroght, that is HISTORY of CREA-
TVRES, HISTORY of MARVAILES, and
HISTORY of ARTS. The frſt of theſe, no doubt
is extánt, and that in good perfection: The two later are handled fo weakely and vnprofitably, as I
am moued to note them as deficient. For I find no
fufficient, or competent Collection of the Woikes   *Hiſtoria*
of Nature, which haue a Digreſſion, and Deflexi-   *Naturæ*
on, from the ordinary courfe of Generations, Pro-   *Errantis.*
ductions, & Motions, whether they be fingularities
of place and region , or the ſtrange euents of time
and chance, or the effects of yet vnknowne proprieties , or the inſtances of exception to generall
kindes: It is true, I finde a number of bookes of
fabulous Experiments, & Secrets, and friuolous Impoſtures for pleafure and ſtrangeneſſe . But a
fubſtantiall and feuere Collection of the HETE-
ROCLITES, or IRREGVLARS of NATVRE,
well examined & defcribed I find not: fpecially not
with due reiection of fables , and popular Errors:
For,

For, as things now are, if an vntruth in Nature bee once on foote, what by reason of the neglect of examination, and countenance of Antiquitie, and what by reason of the vse of the opinion in similitudes, and ornaments of speeche, it is neuer called downe.

The vse of this worke, honoured with a president in *Ariftotle*, is nothing leffe, then to giue contentment to the appetite of Curious and vaine wittes, as the manner of M I R A B I L A R I E S is to doe : But for twoo Reasons, both of greate waight : The one to correct the parcialitie of Axiomes, and Opinions which are commonly framed onely vppon common and familiar examples : The other, becaufe from the Wonders of Nature, is the neereft Intelligence and paffage towardes the Wonders of Arte : For it is no more, but by following, and as it were, hounding Nature in her wandrings, to bee able to leade her afterwardes to the fame place againe. Neyther am I of opinion in this H I S T O R Y of M A R V A I L E S, that fuperftitious Narrations of Sorceries, Witchecraftes, Dreames, Diuinations, and the like, where there is an affurance, and cleere euidence of the fact, be altogether excluded. For it is not yet knowne in what cafes, and how farre, effectes attributed to fuperftition, do participate of Naturall caufes : and therefore how-foeuer the practife of fuch things is to bee condemned, yet from the Speculation and

consideration

sideration of them, light may be taken, not onely for the difcerning of the offences, but for the further difclosing of Nature : Neither ought a Man to make fcruple of entring into thefe things for inquifition of truth, as your Maieftie hath fhewed in your owne example : who with the two cleere eyes of Religion and naturall Philofophy, haue looked deepely and wifely into thefe fhadowes, and yet proued your felfe to be of the Nature of the Sunne, which paffeth through pollutions, and it felfe remaines as pure as before. But this I hold fit, that thefe Narrations, which haue mixture with fuperftition, be forted by themfelues, and not to be mingled with the Narrations, which are meerely and fincerely naturall. But as for the Narrations touching the Prodigies and Miracles of Religions, they are either not true, or not Naturall ; and therefore impertinent for the Storie of Nature.

For HISTORY of NATVRE *Hiftoria* WROVGHT, or MECHANICALL, *Mechani-* I finde fome Collections Made of Agriculture, *ca.* and likewife of Manuall Arts, but commonly with a reiection of experiments familiar and vulgar. For it is efteemed a kinde of difhonour vnto Learning, to defcend to enquirie or Meditation vppon Matters Mechanicall; except they bee fuch as may bee thought fecrets, rarities, and fpeciall fubtilties : which humour of vaine, and

C c   fupercilious

supercilious Arrogancie, is iuftly derided in *Plato* :
where hee brings in *Hippias* a vanting *Sophift*, dif-
puting with *Socrates* a true and vnfained inqui-
fitor of truth ; where the fubieĉt beeing touching
beautie, *Socrates*, after his wandring manner of In-
duĉtions, put firft an example of a faire Virgine, and
then of a faire Horfe, and then of a faire pot well
glazed, whereat *Hippias* was offended, and faid;
More then for curtefies fake, hee did thinke much to dif-
pute with any , that did alledge fuch bafe and Sordide
inftances, whereunto *Socrates* anfwereth ; *you haue*
reafon, and it becomes you well , beeing a man fo
trimme in your vefliments , &c. and fo goeth on
in an Ironie. But the truth is , they bee not the
higheft inftances, that giue the fecureft information ; as may bee well expreffed in the tale fo common of the *Philofopher*, that while he gazed vp-
wardes to the Starres, fell into the water : for if
hee had looked downe hee might haue feene the
Starres in the water, but looking aloft hee coulde
not fee the water in the Starres : So it commeth
often to paffe, that meane and fmall things difcouer great, better then great can difcouer the fmall:
and therefore *Ariftotle* noteth well, *that the nature*
of euery thing is beft feene in his fmalleft portions, and
for that caufe hee enquireth the nature of a Common wealth , firft in a Family, and the Simple
Coniugatiōs of Man and Wife; Parent, and Child,
Maifter and Seruant, which are in euery Cottage;

Euen

Euen so likewise the nature of this great Citie of the world and the policie thereof, must bee first sought in meane concordances, and small portions : So we see how that secret of Nature, of the turning of Iron, touched with the Loadestone, towardes the North, was found out in needels of Iron, not in barres of Iron.

But if my iudgement bee of any waight, the vse of HISTORIE MECHANICAL , is of all others the most radicall, and fundamentall towardes Naturall Philosophie, such Naturall Philosophie, as shall not vanish in the fume of subtile, sublime, or delectable speculation , but such as shall bee operatiue to the endowment, and benefit of Mans life : for it will not onely minister and suggest for the present, Many ingenious practizes in all trades, by a connexion and transferring of the obseruations of one Arte , to the vse of another, when the experiences of seuerall misteries shall fall vnder the consideration of one mans minde : But furder, it will giue a more true , and reall illumination concerning Causes and Axiomes, then is hetherto attained. For like as a Mans disposition is neuer well knowen, till hee be crossed, nor *Proteus* euer chaunged shapes , till hee was straightened and held fast : so the passages and variations of Nature cannot appeare so fully in the libertie of Nature, as in the trialls and vexations of Art.

FOr CIVILE HISTORY, it is of three kinds, not vnfitly to be compared with the three kinds of Pictures or Images : for of Pictures or Images, wee fee fome are Vnfinifhed, fome are parfite, and fome are defaced : So of Hiftories, wee may finde three kindes, MEMORIALLS, PARFITE HISTORIES, and ANTIQVITIES : for MEMORIALLS are Hiftorie vnfinifhed, or the firft, or rough draughts of Hiftorie, and ANTIQVITIES are Hiftorie defaced, or fome remnants of Hiftory, which haue cafually efcaped the fhipwrack of time.

MEMORIALLS or PREPARATORY HISTORY are of 2 forts, wherof the one may be tearmed COMMENTARIES, & the other REGISTERS : COMMENTARIES are they which fet downe a continuance of the naked euéts & actiõs, without the motiues or defignes, the counfells, the fpeeches, the pretexts, the occafions, and other paffages of action : for this is the true nature of a commentarie (though *Cafar* in modeftie mixt with greatneffe, did for his pleafure apply the name of a commentarie to the beft Hiftorie of the world) REGISTERS are collectiõs of Publique Acts, as Decrees of counfell, Iudiciall proceedings, Declarations and Letters of eftate, Orations, and the like, without a perfect continuance, or contexture of the threed of the Narration.

ANTIQVI.

ANTIQVITIES, or Remnants of Histo-
ry, are, as was saide, *tanquam Tabula Naufragij*, when
induſtrious perſons by an exact and ſcrupulous diligence and obſeruation, out of Monumēts, Names,
Wordes, Prouerbes, Traditions, Priuate Recordes,
and Euidences, Fragments of ſtories, Paſſages of
Bookes, that concerne not ſtorie, and the like, doe
ſaue and recouer ſomewhat from the deluge of
time.

In theſe kindes of vnperfect Hiſtories *I* doe aſ-
ſigne no deficience, for they are *tanquam imperfectè
Miſſa,* and therefore any deficience in them is but
their nature. As for the Corruptiōs and Mothes of
Hiſtorie, which are *Epitomes,* the vſe of them deſerueth to be baniſht, as all men of ſound Iudgement haue confeſſed, as thoſe that haue fretted
and corroded the ſound bodies of many excellent
Hiſtories, and wrought them into baſe and vnprofitable dregges.

HISTORY which may be called IVST
and PARFITE Hiſtorie, is of three kinds, according to the object which it propoundeth, or pretendeth to repreſent: for it either repreſēteth a TIME,
or a PERSON, or an ACTION. The firſt we
call CHRONICLES, The ſecond LIVES,
and the third NARRATIONS, or RELA-
TIONS. Of theſe although the firſt bee the
moſt compleate and abſolute kinde of Hiſtorie, and
hath moſt eſtimation and glory : yet the ſecond excelleth it in profit and vſe, and the third in veritie

& finceritie. For HISTORY of TIMES reprefen-
teth the magnitude of Actions, &the publique faces
and deportmēts of perfons, & paffeth ouer in filence
the fmaller paffages and Motions of men and Matters.　But fuch beeing the workemanfhip of God,
as he doth hang the greateft waight vpon the fmalleft Wyars, *Maxima è Minimis fufpēdēs*, it comes therfore to paffe, that fuch Hiftories doe rather fet forth
the pompe of bufines, then the true and inward re-
forts thereof. But *Liues* if they be well written, pro-
pounding to themfelues a perfon to reprefent, in
whom actions both greater and fmaller, publique &
priuate haue a commixture; muft of neceffitie containe a more true, natiue, and liuely reprefentation:
So againe Narrations, and Relations of actions as
the War of *Peloponnefus*, the Expeditiō of *Cyrus Minor*, the Confpiracie of *Catiline*, cannot but be more
purely and exactly true, then HISTORIES of
TIMES, becaufe they may choofe an argument
comprehenfible within the notice and inftructions
of the Writer: whereas he that undertaketh the fto-
ry of a time, fpecially of any length, cannot but meet
with many blankes, and fpaces, which hee muft be
forced to fill vp, out of his own wit and coniecture.
　　For the HISTORY of TIMES, (I meane
of ciuil Hiftory,) the prouidence of God hath made
the diftribution: for it hath pleafed God to ordaine
and illuftrate two exemplar States of the worlde,
for Armes, learning, Morall Vertue, Policie,
and Lawes. The STATE of GRECIA, and
　　　　　　　　　　　　　　　　　　　　the

the STATE of *ROME* : The Histories whereof occupying the MIDDLE PART of time,
haue more auncient to them, Histories which may
by one common name, be tearmed the ANTI-
QVITIES of the WORLD; and after
them , Histories which may bee likewise called by the name of MODERNE HISTO-
RIE.

Nowe to speake of the deficiences : As to the
HEATHEN ANTIQVITIES of the
world, it is in vaine to note them for deficient : deficient they are no doubt , consisting most of fables
and fragments : but the deficience cannot bee hol-
pen : for Antiquitie is like fame , *Caput inter nubila
condit* , her head is muffled from our sight: For the
HISTORIE of the EXEMPLAR
STATES , it is extant in good perfection.
Not but I could wish there were a perfect Course
of Historie for *Grecia* from *Theseus* to *Philopæmen* ,
(what time the affaires of *Grecia* drowned and
extinguished in the affaires of *Rome*) and for *Rome,*
from *Romulus* to *Iustinianus* , who may be truly
saide to be *Vltimus Romanorum.* In which sequences
of storie the Text of *Thucidides* and *Xenophon* in the
one, & the Texts of *Liuius, Polybius, Salustius, Cæsar,
Appianus, Tacitus, Herodianus* in the other to be kept
intyre without any diminutiõ at all, and onely to be
supplied and continued. But this is Matter of Magnificence, rather to be commended then required:
and

and wee fpeake nowe of parts of Learning fupple-
mentall, and not of fupererogation.

But for MODERNE HISTORIES, where-
of there are fome fewe verie worthy, but the greater part beneath Mediocritie, leauing the care of
forreyne ftories to forreyne States, becaufe I will
not bee *Curiofus in aliena Republica*, I cannot faile
to reprefent to your Maieftie, the vnworthineffe
of the Hiftorie of *Englande* in the Maine continuance thereof, and the partialitie, and obliquitie
of that of *Scotland*, in the lateft and largeft Author
that I haue feene; fuppofing that it would be honour for your Maieftie, and a worke very memorable, if this Iland of great *Brittanie*, as it is now ioyned in Monarchie for the ages to come: So were
ioyned in one Hiftorie for the times paffed, after
the manner of the facred Hiftorie, which draweth
downe the ftorie of the Tenne Tribes, and of the
Two Tribes, as Twinnes together. And if it fhall
feeme that the greatneffe of this worke may make
it leffe exactly performed, there is an excellent Periode of a much fmaller compaffe of time, as to the
ftorie of *England* that is to fay, from the Vniting of
the Rofes, to the Vniting of the Kingdomes: a Por-
tió of time wberin, to my vnderftanding, there hath
bin the rareft varieties, that in like number of fuccef-
fiós of any hereditary Monarchie hath bin known:
For it beginneth with the mixt Adeption of a
Crowne, by Armes and Tytle: An entry by Bat-
taile, an Eftablifhment by Mariage: and therefore
times

times anſwerable, like waters after a tempeſt, full of working and ſwelling, though without extreamitie of Storme; but well paſſed through by the wiſedome of the *Pylote*, being one of the moſt ſufficient kinges of all the number. Then followeth the Raigne of a King, whoſe actions howſoeuer conducted had much intermixture with the affaires of *Europe* : balancing and inclyning them variably, in whoſe time alſo beganne that great alteration in the State Eccleſiaſticall, an action which ſeldome commeth vppon the Stage : Then the Raigne of a Minor, then an offer of an vſurpation, (though it was but as *Febris Ephemera*). Then the Raigne of a Queene Matched with a Forreyner : Then of a Queene that liued ſolitary, and vnmarried, and yet her gouernment ſo maſculine, as it had greater impreſſion, and operation vppon the States abroad, then it any waies receiued from thence; And now laſt, this moſt happie, and glorious euent, that this Iland of *Brittany* deuided from all the world, ſhould bee vnited in it ſelfe ; And that Oracle of Reſt giuen to *Aeneas, Antiquam enquirite Matrem,* ſhould nowe bee performed and fulfilled vpon the Nations of *England* and *Scotland,* being now revnited in the auncient Mother name of *Brittany*, as a full periode of all inſtabilitie & peregrinations : So that as it commeth to paſſe in Maſsiue bodies, that they haue certaine trepidations and wauerings before they fixe and ſettle : So it ſeemeth, that

D d   by

by the prouidence of God, this Monarchy before
it was to settle in your Maieftie, and your generati-
ons , (in which I hope it is nowe eftablifhed for
euer,) it had thefe prelufiue changes and varieties.

For LIVES, I doe finde ftrange that thefe
times haue fo litle efteemed the vertues of the times,
as that the Writings of liues fhould be no more frequent. For although there be not many foueraigne
Princes or abfolute cõmanders, and that States are
moft collected into Monarchies; yet are there many
worthy perfonages, that deferue better then difperfed report, or barren *Elogies* : For herein the Inuention of one of the late Poets is proper, and doth well
inrich the auncient fiction; for he faineth, that at the
end of the threed or *Webbe* of euery mans life, there
was a little *Medall* containing the *Perfons* name,
and that *Time* waited vpon the fheeres, and affoone
as the threed was cut, caught the Medalls, and carried them to the Riuer of *Lethe*, and about the banke
there were many Birds flying vp and downe, that
would get the Medals and carry them in their Beke
a little while, and then let them fall into the Riuer.
Onely there were a fewe Swannes , which if they
got a Name , would carrie it to a Temple, where
it was confecrate . And although many men
more mortall in their affections , then in their bodies , doe efteeme defire of name and memory, but
as a vanitie and ventofitie ;

Animi nil magnæ laudis egentes :

Which

Which opinion commeth from that Root, *Non pri-
us laudes contempsimus, quam laudanda facere desivi-
mus*: yet that will not alter *Salomons* iudgement, *Me-
moria Iusti cum laudibus, at impioru nome putrescet*: The
one flourisheth, the other either côsumeth to preset
obliuion, or turneth to an ill odor: And therefore in
that stile or addition, which is & hath bin long well
receiued, and brought in vse, *Fælicis memoriæ, piæ memoriæ, bonæ memoriæ*, we do acknowledge that which
Cicero saith, borrowing it frô *Demosthenes*, that *Bona
Fama propria possessio defunctorum*, which possession
I cânot but note, that in our times it lieth much wast
and that therein there is a Deficience.

For NARRATIONS and RELATIONS
of particular actions, there were also to be wished a
greater diligence therein, for there is no great action but hath some good penne which attends it.
And because it is an abilitie not common to Write
a good History, as may well appeare by the small
number of them: yet if particularitie of actions memorable, were but tolerably reported as they passe,
the compiling of a complete HISTORIE of
TIMES mought be the better expected, when a
Writer should arise that were fit for it: for the collection of such relations mought be as a Nursery gardein, whereby to plant a faire and stately gardein,
when time should serue.

There is yet another portion of Historie which
Cornelius Tacitus maketh, which is not to be forgottê

specially

ſpecially with that application, which hee accou-
pleth it withal, A N N A L S, and I O V R N A L S,
appropriating to the former, Matters of eſtate, and
to the later, Acts, and Accidents of a meaner Nature. For giuing but a touch of certaine Magnificent Buildings, he addeth, *Cum ex dignitate populi Ro-
mani repertum ſit, res illuſtres annalibus, talia diurnis
vrbis Actis mandare*. So as there is a kinde of contemplatiue Heraldry, as well as Ciuill. And as nothing doth derogate from the dignitie of a ſtate
more then confuſion of degrees : So it doth not a
little imbaſe the Authoritie of an Hiſtorie, to intermingle matters of triumph, or matters of ceremo-
ny, or matters of Noueltie, with matters of State ;
But the vſe of a I o v r n a l l hath not onely been
in the Hiſtorie of Time, but likewiſe in the Hiſtorie
of Perſós, and chiefely of actions; for Princes in ancient time had vpon point of honour and policie
both, Iournalls kept, what paſſed day by day : for
we ſee the Chronicle which was red before *Ahaſſu-
erus*, when he could not take reſt, contained matter
of affaires indeede, but ſuch as had paſſed in his own
time, and very lately before ; But the I o v r n a l l
of *Alexanders* houſe expreſſed euery ſmall particularitie, euen concerning his Perſon and Court: and it
is yet an vſe wel receiued in enterpriſes memorable,
as expeditions of Warre, Nauigations, and the like,
to keepe *Dyaries* of that which paſſeth continually.
　　I cannot likewiſe bee ignorant of a forme of
Writing, which ſome graue and wiſe men haue
vſed,

vfed, containing a fcattered Hiftory of thofe actions, which they haue thought worthy of memorie, with politique difcourfe and obferuation thereupon; not incorporate into the Hiftory, but feperately, and as the more principall in their intentiō: Which kind of R v m i n a t e d H i s t o r y, I thinke more fit to place amongft Bookes of policie, whereof we fhall hereafter fpeake, then amongft Bookes of Hiftory: for it is the true office of Hiftory to reprefent the e-uents themfelues, together with the counfels, and to leaue the obferuations, and conclufions thereup-on, to the liberty and facultie of euery mans iudge-ment: But Mixtures, are things irregular, whereof no man can define.

So alfo is there another kinde of Hiftory mani-foldly mixt, and that is H i s t o r y of C o s m o-g r a p h y, being compounded of Naturall Hiftory in refpect of the Regions themfelues, of Hiftory ci-uill, in refpect of the Habitations, Regiments, and Manners of the people; and the *Mathematiques* in refpect of the Climats, and configurations towards the Heauens, which part of learning of all others in this latter time hath obtained moft Proficience. For it may be truly affirmed to the honor of thefe times, and in a vertuous emulation with Antiquitie, that this great Building of the world, had neuer *through lights* made in it, till the age of vs and ourfathers: For although they had knowledge of the *Antipodes*:

Nofque vbi primus equis Oriens afflauit anhelis:

Illic

Illic sera rubens accēdit lumina vesper, yet that moughʒ be by demonſtration, and not in faƈt, and if by trauaile, it requireth the voiage but of halfe the Globe. But to circle the Earth, as the heauenly Bodies doe, was not done, nor enterpriſed, till theſe later times: And therefore theſe times may iuſtly beare in their word, not onely *Plus vltrà* in precedence of the ancient *Non vltrà*, and *Imitabile fulmen*, in precedence of the ancient :   *Non imitabile fulmen,*

   Demens qui nymbos et non imitabile fulmen, &c.
But likewiſe,   *Imitabile Cælum:*
in reſpeƈt of the many memorable voyages after the maner of heauē, about the globe of the earth.

And this Proficience in Nauigation, and diſcoueries, may plant alſo an expeƈtation of the furder proficience, and augmentation of all Scyences, becauſe it may ſeeme they are ordained by God to be *Coevalls*, that is, to meete in one Age. For ſo the Prophet *Daniel* ſpeaking of the latter times foretelleth: *Plurimi pertranſibunt, & Multiplex erit Scientia,* as if the openneſſe and through paſſage of the world, and the encreaſe of knowledge were appointed to be in the ſame ages, as we ſee it is already performed in great part, the learning of theſe later times not much giuing place to the former two Periods or Returnes of learning, the one of the Græcians, the other of the Romanes.

Hɪꜱᴛᴏʀʏ Eᴄᴄʟꜱɪᴀꜱᴛɪᴄᴀʟ, receiueth the ſame diuiſions with Hiſtory ciuil; but furder in
the

the proprietie thereof may bee deuided into H I S-
T O R Y of the C H V R C H, by a general name. H I S-
T O R Y of P R O P H E C I E, & H I S T O R I E of P R O-
V I D E N C E : The firſt deſcribeth the times of the
militant Church; whether it be fluctuât, as the Arke
of *Noah*, or moueable, as the Arke in the Wildernes,
or at reſt, as the Arke in the Temple; That is, the ſtate
of the Church in Perſecution, in Remoue, and in
Peace. This part I ought in no ſort to note as deficient, onely I would the vertue and ſinceritie of it,
were according to the Maſſe, and quantitie. But I
am not now in hand with cenſures, but with omiſſions.

The ſecond, which is H I S T O R Y of P R O-
P H E C I E, conſiſteth of two Relatiues, the Prophecie, and the accompliſhment: and therefore the
nature of ſuch a worke ought to be, that euery prophecie of the Scripture be ſorted with the euent ful-
filling the ſame, throughout the ages of the world,
both for the better confirmation of faith, and for
the better illumination of the Church, touching
thoſe parts of Prophecies, which are yet vnfulfilled:
allowing neuertheleſſe that Latitude, which is a-
greable, and familiar vnto diuine Prophecies, being of the nature of their Author, with whom a
thouſande yeares are but as one day, and therefore are not fulfilled punctually, at once, but
haue ſpringing and germinant accompliſhment
throughout many ages, though the height or
fulneſſe of them may referre to ſome one age:
This

Hiſtoria Propheti-ca.

This is a worke which I finde deficient, but is to bee done with wiſedom, ſobrietie, and reuerence, or not at all.

The third, which is H I S T O R Y of P R O V I-D E N C E, containeth that excellēt correſpondence, which is betweene Gods reuealéd will, and his ſe-cret will: which though it be ſo obſcure, as for the moſt part it is not legible to the Naturall Man; no, nor many times to thoſe that behold it from the Ta-bernacle: yet at ſome times it pleaſeth God for our better eſtabliſhment, and the confuting of thoſe which are as without God in the world; to write it in ſuch Text and Capitall Letters, that, as the Pro-phet ſaith, *He that runneth by, may read it*: that is, meere ſenſual perſons, which haſten by Gods iudgements, and neuer bend or fixe their cogitations vpon them, are neuertheleſſe in their paſſage and race vrged to diſcerne it. Such are the notable euents and exam-ples of Gods iudgements, chaſtizements, deliue-rances and bleſſings: And this is a work which hath paſſed through the labour of many, and therefore I cannot preſent as omitted.

There are alſo other parts of learning which are A P P E N D I C E S to H I S T O R Y, for al the exterior proceedings of man conſiſt of Wordes and Deeds: whereof Hiſtory doth properly receiue, and retaine in Memory the Deedes, and if Wordes, yet but as Inducements and paſſages to Deedes: So are there other Bookes and Writings, which are appropriat to the cuſtodie, and receite of Wordes onely: which

likewiſe

likewife are of three forts. O R A T I O N S, L E T-
T E R S, & B R I E F E S P E E C H E S, or S A Y-
I N G S: O R A T I O N S are pleadings, fpeeches
of counfell ; Laudatiues, Inuectiues, Apologies,
Reprehenfions ; Orations of Formalitie , or Ceremonie, and the like : Letters are according to all
the varietie of occafions; Aduertifments, Aduifes,
Directions, Propofitions, Peticions, Commenda-
torie, Expoftulatorie, Satiffactorie, of complement,
of Pleafure, of Difcourfe, and all other paffages
of Action. And fuch as are written from wife men,
are, of all the words of Man, in my iudgement the
beft, for they are more Naturall then Orations, and
publike fpeeches, & more aduifed then coferences,
or prefent fpeeches : So againe Letters of Affaires
from fuch as Manage them, or are priuie to them,
are of all others the beft inftructions for Hiftory, and
to a diligent reader, the beft Hiftories in themfelues.
For A P O T H E G M E S : It is a great loffe of that
Booke of *Cafars* ; For as his Hiftory, and thofe fewe
Letters of his which wee haue, and thofe Apothegmes which were of his owne, excell all mens elfe:
So I fuppofe would his collection of A P O T H E-
G M E S haue done; For as for thofe which are collected by others , either I haue no taft in fuch Matters, or elfe their choice hath not beene happie. But
vpon thefe three kindes of Writings I doe not in-
fift, becaufe I haue no deficieces to propound concerning them,　　　　E e

　　　　　　　　　　　　　　　　Thus

Thus much therefore concerning Hiſtory, which is that part of learning, which anſwereth to one of the Celles, *Domiciles*, or offices of the Mind of Man; which is that of the Memorie.

POESIE is a part of Learning in meaſure of words for the moſt part reſtrained : but in all other points extreamely licenſed : and doth truly referre to the Imagination : which beeing not tyed to the Lawes of Matter ; may at pleaſure ioyne that which Nature hath ſeuered: & ſeuer that which Nature hath ioyned , and ſo make vnlawfull Matches & diuorſes of things: *Pictoribus atque Poetis &c.* It is taken in two ſenſes in reſpect of Wordes or Matter; In the firſt ſenſe it is but a *Character* of ſtile, and belongeth to Arts of ſpeeche , and is not pertinent for the preſent. In the later, it is (as hath beene ſaide) one of the principalll Portions of learning : and is nothing elſe but FAINED HISTORY, which may be ſtiled as well in Proſe as in Verſe.

The vſe of this FAINED HISTORIE, hath beene to giue ſome ſhadowe of ſatiſfaction to the minde of Man in thoſe points, wherein the Nature of things doth denie it, the world being in proportion inferiour to the ſoule : by reaſon whereof there is agreeable to the ſpirit of Man, a more ample Greatneſſe , a more exact Goodneſſe ; and a more abſolute varietie then can bee found in the Nature of things . Therefore , becauſe the Acts

or

or Euents of *true Hiſtorie*, haue not that Magnitude, which ſatisfieth the minde of Man, *Poeſie* faineth Acts and Euents Greater and more Heroicall; becauſe *true Hiſtorie* propoundeth the ſucceſſes and iſſues of actions, not ſo agreable to the merits of Vertue and Vice, therefore *Poeſie* faines them more iuſt in Retribution, and more according to Reuealed Prouidence, becauſe *true Hiſtorie* repreſenteth Actions and Euents, more ordinarie and leſſe interchanged, therefore *Poeſie* endueth them with more Rareneſſe, and more vnexpected, and alternatiue Variations. So as it appeareth that *Poeſie* ſerueth and conferreth to Magnanimitie, Moralitie, and to delectation. And therefore it was euer thought to haue ſome participation of diuineſſe, becauſe it doth raiſe and erect the Minde, by ſubmitting the ſhewes of things to the deſires of the Mind; whereas reaſon doth buckle and bowe the Mind vnto the Nature of things. And we ſee that by theſe inſinuations and congruities with mans Nature and pleaſure, ioyned alſo with the agreement and conſort it hath with Muſicke, it hath had acceſſe and eſtimation in rude times, and barbarous Regions, where other learning ſtoode excluded.

The diuiſiõ of Poeſie which is apteſt in the proprietie therof (beſides thoſe diuiſiõs which are cõmon vnto it with hiſtory: as fained Chronicles, fained liues, & the Appẽdices of Hiſtory, as fained Epiſtles, fained Orations, and the reſt) is into POESIE

NARRA-

NARRATIVE, REPRESENTATIVE, and ALLVSIVE. The NARRATIVE is a meere imitation of Hiftory with the exceffes before remembred; Choofing for fubiect comonly Warrs, and Loue; rarely State, and fometimes Pleafure or Mirth. REPRESENTATIVE is as a vifible Hiftory, and is an Image of Actions as if they were prefent, as Hiftory is of actions in nature as they are, that is paft; ALLVSIVE or PARABOLI-CALL, is a NARRTION applied onely to expreffe fome fpeciall purpofe or conceit. Which later kind of Parabolical wifedome was much more in vfe in the ancient times, as by the Fables of *Ae-fope*, and the briefe fentences of the feuen, and the *vfe* of *Hieroglyphikes* may appeare. And the caufe was for that it was then of neceffirie to expreffe any point of reafon, which was more fharpe or fubtile then the vulgar in that maner, beeaufe men in thofe times wanted both varietie of examples, and fubtil-tie of conceit : And as *Hierogliphikes* were before Letters, fo parables were before arguments : And neuerthelelfe now and at all times they doe retaine much life and vigor, becaufe reafon cannot bee fo fenfible, nor examples fo fit.

But there remaineth yet another vfe of POESY PARABOLICAL, oppofite to that which we laft mentioned: for that tendeth to demonftrate, and illuftrate that which is taught or deliuered, and this other to retire and obfcure it : That is when the Se-crets and Mifteries of Religion, Pollicy, or Philofo-phy,

phy, are inuolued in Fables or Parables. Of this in
diuine Poefie, wee fee the vfe is authorifed. In Heathen Poefie, wee fee the expofition of Fables doth
fall out fometimes with great felicitie, as in the Fable that the Gyants beeing ouerthrowne in their
warre againft the Gods, the Earth their mother in
reuenge thereof brought forth Fame.

Illam terra Parens ira irritata Deorū, (Progenuit.

Extremam,vt perhibent,Cœo Enceladoque Sororem
expounded that when Princes & Monarchies haue
fuppreffed actuall and open Rebels,then the malignitie of people, which is the mother of Rebellion,
doth bring forth Libels & flanders, and taxatiōs of
the ftates,which is of the fame kind with Rebellion,
but more Feminine: So in the Fable that the reft of
the Gods hauing confpired to binde *Iupiter*, *Pallas*
called *Briareus* with his hundreth hands to his aide,
expounded, that Monarchies neede not feare any
courbing of their abfoluteneffe by Mightie Subiects,as long as by wifedome they keepe the hearts
of the people, who will be fure to come in on their
fide:So in the fable,that *Achilles* was brought vp vnder *Chyron* the *Centaure*,who was part a man, & part
a beaft,expounded Ingenuoufly, but corruptly by
Machiauell, that it belongeth to the education and
difcipline of Princes, to knowe as well how to play
the part of the Lyon, in violence, and the Foxe in
guile,as of the man in vertue and Iuftice.Neuerthe-
leffe in many the like incounters, I doe rather think
that the fable was firft, and the expofition deuifed,

then

then that the Morall was firſt,& thereupon the fable framed. For I finde it was an auncient vanitie, in *Chriſippus*, that troubled himſelfe with great contention to faſten the aſſertions of the *Stoicks* vpon fiₒctions of the ancient Poets : But yet that all the Fables and fiₒctions of the Poets, were but pleaſure and not figure., I interpoſe no opinion. Surely of thoſe Poets which are now extant, euen *Homer* himſelfe, (notwithſtanding he was made a kinde of Scripture,by the later Schooles of the Grecians) yet I ſhould without any difficultie pronounce, that his Fables had no ſuch inwardneſſe in his owne meaning : But what they might haue, vpon a more originall tradition, is not eaſie to affirme,for he was not the inuentor of many of them. In this third part of Learning which is Poeſie, I can report no deficience. For being as a plant that commeth of the luſt of the earth,without a formall ſeede, it hath ſprung vp,and ſpread abroad, more then any other kinde: But to aſcribe vnto it that which is due for the expreſſing of affeₒctions, paſſions, corruptions and cuſtomes,we are beholding to Poets, more thē to the Philoſophers workes, and for wit and eloquence not much leſſe then to Orators harangues. But it is not good to ſtay too long in the Theater: let vs now paſſe on to the iudicial Place or Pallace of the Mind, which we are to approach and view, with more reuerence and attention.

T he knowledge of Man is as the waters, ſome
                                                    deſcending

descending from aboue, and some springing from
beneath, the one informed by the light of Nature,
the other inspired by diuine reuelation. The light of
Nature consisteth, in the Notions of the minde, and
the Reports of the Sences, for as for knowledge
which Man receiueth by teaching, it is Cumulatiue,
and not Originall, as in a water, that besides his own
spring-heade is fedde with other Springs and
Streames. So then according to these two differing
Illuminations, or Originals, Knowledge is first of al
deuided into DIVINITIE, and PHILOSO-
PHIE.

In PHILOSOPHY, the contemplations of
Man doe either penetrate *vnto God*, or are circum-
ferred to *Nature*, or are reflected or reuerted vpon
himselfe. Out of which seuerall inquiries, there doe
arise three knowledges, DIVINE PHILO-
SOPHY, NATVRAL PHILOSOPHY,
and HVMANE PHILOSOPHY, or
HVMANITIE. For all things are marked and
stamped with this triple Character of the power of
God, the difference of Nature, and the vse of Man.
But because the distributions and partitions of
knowledge, are not like seuerall lines, that
meete in one Angle, and so touch but in a point,
but are like branches of a tree, that meete in a steme;
which hath a dimension and quantitie of en-
tyrenes and continuance, before it come to discon-
tinue & break it self into Armes and boughes, therfore

fore it is good, before wee enter into the former diftribution, to erect & conftitute one vniuerfal Science by the name of P H I L O S O P H I A P R I-M A, P R I M I T I V E or S V M M A R I E P H I-L O S O P H I E, as the Maine and common way, before we come where the waies part, and deuide themfelues, which Sciéce whether I fhould report as deficient or noe, I ftand doubtfull. For I finde a certaine Rapfodie of *Naturall Theologie,* and of diuers parts of *Logicke* : And of that part of *Naturall Philofophie*, which concerneth the *Principles*, and of that other part of *Naturall Philofophy*, which concerneth the *Soule* or *Spirit*, all thefe ftrangely commixed and confufed : but being examined it feemeth to mee rather a depredation of other Sciences, aduanced and exalted vnto fome height of tearmes, then any thing folide or fubftantiue of it felfe. Neuerthelefle I cannot bee ignorant of the diftinction which is currant, that the fame things are handled but in feuerall refpects : as for example that Logicke confidereth of many things as they are in Notion: & this Philofophy. as they are in Nature: the one in, Apparance, the other in Exiftence : But I finde this difference better made then purfued ; For if they had confidered *Quantitie*, *Similitude*, *Diuerfitie*, and the reft of thofe *Externe Characters* of things, as *Philofophers*, and in Nature: their inquiries muft

of

of force haue beene of a farre other kinde then
they are. For doth anie of them in handeling
Quantitie, fpeake of the force of vnion, how, and
how farre it multiplieth vertue? Doth any giue the
reafon, why fome things in Nature are fo common
and in fo great Mafse, and others fo rare, and in fo
fmall quantitie? Doth anie in handling Similitude and Diuerfitie, affigne the caufe why Iron
fhould not mooue to Iron, which is more like, but
mooue to the Loadeftone, which is leffe like?
why in all Diuerfities of things there fhould bee
certaine Participles in Nature, which are almoft
ambiguous, to which kinde they fhould bee referred? But there is a meere and deepe filence, touching the Nature and operation of thofe *Common
adiuncts* of things, as in Nature; and onely a re-
fuming and repeating of the force and vfe of them,
in fpeeche or argument. Therefore becaufe in a
Wryting of this Nature, I auoyde all fubtilitie:
my meaning touching this Originall or vniuerfall
Philofophie, is thus in a plaine and groffe defcription by Negatiue: *That it bee a Receptacle for all
fuch profitable obferuations and Axioms, as fall not
within the compaffe of any of the fpeciall parts of Philofophie, or Sciences; but are more common, and of a
higher ftage.*

Now that there are manie of that kinde neede
not bee doubted: for example; Is not the rule:
Si inæqualibus æqualia addas, omnia erunt inæqualia.

F f   And

An Axiome afwell of *Iustice*, as of the Mathematiques ? And is there not a true coincidence betweene commutatiue and diftributiue *Iustice*, and Arithmeticall and Geometricall proportion? Is not that other rule, *Quæ in eodem tertio conveniunt, & inter se conveniunt*, a Rule taken from the Mathematiques, but fo potent in Logicke as all Syllogifmes are built vppon it ? Is not the obferuation, *Omnia mutantur, nil interit*, a contemplation in Philofophie thus, that the *Quantum* of Nature is eternall, *In* Naturall Theologie thus, That it requireth the fame Omnipotencie to make fomewhat Nothing, which at the firft made nothing fomewhat ? according to the Scripture, *Didici quod omnia opera quæ fecit Deus, perfeuerent in perpetuum, non possumus eis quicquam addere, nec auferre*. Is not the ground which *Machiavill* wifely and largely difcourfeth concerning Gouernments, That the way to eftablifh and preferue them, is to reduce them *ad Principia*; a rule in Religion and Nature, afwell as in Ciuill adminiftration ? was not the *Perfian* Magicke a reduction or correfpondence of the Principles & Architectures of Nature, to the rules and policie of Gouernments ? Is not the precept of a Mufitian, to fall from a difcord or harfh accord, vpon a concord, or fweete accord, alike true in affection ? Is not the Trope of Muficke, to auoyde or flyde from the clofe or Cadence, common with the Trope of Rhetoricke of deceyuing expectation ? Is not the delight

light of the Quauering vppon a ftoppe in Muficke, the fame with the playing of Light vppen
the water ?

   - *Splendet tremulo fub Lumine Pontus.*

   Are not the Organs of the fences of one kinde
with the Organs of Reflexion, the Eye with a glaffe,
the Eare with a Caue or Straight determined and
bounded ? Neither are thefe onely fimilitudes, as
men of narrowe obferuation may conceyue them
to bee; but the fame footeftcppes of Nature, trea-
ding or printing vppon feuerall fubiects or Matters.
This Science therefore (as I vnderftand it) I may *Philofo-*
iuftlie reporte as deficient ; for I fee fometimes the *phia pri-*
profounder fort of wittes, in handeling fome parti- *ma, fiue de*
culer argument, will nowe and then drawe a Buc- *Fontibus*
ket of Water out of this well, for their prefent vfe: *Scientia-*
But the fpringhead thereof feemeth to mee, not to *rum.*
haue beene vifited; beeing of fo excellent vfe, both
for the difclofing of Nature, and the abridgement
of Art.

   This fcience beeing therefore firft placed as
a common parent, like vnto *Berecinthia*, which
had fo much Heauenlie yffue, *Omnes Cælicolas, om-*
nes fupera alta tenentes ; wee may returne to the former diftribution of the three philofophies; Diuine,
Naturall, and Humane.   And as concerning D i-
vine Philosophie, or Natvrall
Theologie,   It is that knowledge or Rudiment of knowledge concerning G o d, which may
be obtained by the contemplation of his Creatures:

              F f 2   which

which knowledge may bee truely tearmed Diuine, in respect of the obiect; and Naturall in respect of the Light.   The boundes of this knowledge are, that it sufficeth to conuince Atheisme; but not to informe Religion :   And therefore there was neuer Miracle wrought by God to conuert an A-theist, bycause the light of Nature might haue ledde him to confesse a God :   But Miracles haue beene wrought to conuert Idolaters, and the superstitious, because no light of Nature extendeth to declare the will and true worship of God.   For as all works do shewe forth the power and skill of the workeman, and not his Image :   So it is of the works of God; which doe shew the Omnipotencie and wisedome of the Maker, but not his Image :   And therefore therein the Heathen opinion differeth from the Sacred truth : For they supposed the world to bee the Image of God, & Man to be an extract or compendious Image of the world : But the Scriptures neuer vouch-safe to attribute to the world that honour as to bee the Image of God :   But onely *The worke of his hands* ,   Neither do they speake of any other Image of God, but Man: wherfore by the contemplation of Nature ,   to induce and inforce the acknowledgement of God , and to demonstrate his power, prouidence, and goodnesse, is an excellent argument, and hath beene excellently handled by diuerse.   But on the other side, out of the contemplation of Nature , or ground of humane knowe-

ledges,

ledges to induce any veritie, or perſwaſion concerning the points of Faith, is in my iudgement, not
ſafe : *Da fidei, quæ fidei ſunt.* For the Heathen themſelues conclude as much in that excellent and Diuine fable of the Golden Chayne : *That men and
Gods were not able to draw Iupiter down to the Earth, but
contrariwiſe, Iupiter was able to draw them vp to Heauen.*
So as wee ought not to attempt to drawe downe or
ſubmitte the Myſteries of G o d to our Reaſon: but
contrarywiſe, to raiſe and aduance our Reaſon to
the Diuine Truthe. So as in this parte of knowledge, touching Diuine Philoſophie : I am ſo
farre from noting any deficience, as I rather note
an exceſſe : wherevnto I haue digreſſed, becauſe
of the extreame preiudice, which both Religion
and Philoſophie hath receiued, and may receiue
by beeing commixed togither ; as that which vndoubtedly will make an Hereticall Religion ; and
an *I*maginarie and fabulous Philoſophie.

Otherwiſe it is of the Nature of Angells
and Spirits, which is an Appendix of Theologie,
both Diuine and Naturall; and is neither inſcrutable nor interdicted : For although the Scripture
ſaith, *Lette no man deceyue you in Sublime diſcourſe
touching the worſhip of Angells, preſſing into that hee
knoweth not, &c.* Yet notwithſtanding if you obſerue well that precept, it may appeare thereby,
that there bee two things onely forbidden, Ado-

ration

ration of them, and Opinion Fantafticall of them,
eyther to extoll them , further then appertaineth to
the degree of a Creature; or to extoll a mans knowledge of them, further then hee hath ground. But
the fober and grounded inquirie which may arife
out of the paffages of holie Scriptures, or out of the
gradacions of Nature is not reftrained : So of degenerate and reuolted fpirites; the conuerfing with
them, or the imployement of them is prohibited.;
much more any veneration towards them. But the
contemplacion or fcience of their Nature , their
power, their illufions, either by Scripture or reafon,
is a part of fpirituall Wifedome. For fo the Apoftle
faieth, *Wee are not ignorant of his Stratagems* : And
it is no more vnlawfull to enquire the Nature of
euill fpirites , then to enquire the force of poyfons
in Nature, or the Nature of finne and vice in Moralitie; But this parte touching Angells and Spi-
rites, I cannot note as deficient , for many haue oc-
cupyed themfelues in it : I may rather challenge
it in manie of the Wryters thereof, as fabulous and
fantafticall.

Leauing therefore DIVINE PHILOSOPHY,
or NATVRALL THEOLOGIE, (not DIVI-
NITIE, or INSPIRED THEOLOGIE, which
wee referue for the laft of all , as the Hauen and Sab-
bath of all Mans contemplations) wee will nowe
proceede to NATVRALL PHILOSOPHIE : If
then it bee true that *Democritus* fayde, *That the*
truth

truthe of Nature lyeth hydde in certaine deepe Mynes
and Caues ;   And if it bee true likewife, that the Al-
chymists doe fo much inculcate , That Vulcan is a
fecond Nature, and imitateth that dexterouflie and
compendiouflie , which Nature worketh by ambages,& length of time, It were good to deuide Naturall Phylofophie into the Myne and the Fornace,
and to make two profeffions or occupations of Naturall Philofophers , fome to bee Pionners , and
fome Smythes, fome to digge , and fome to refine,
and Hammer :   And furely I doe beft allowe of a
diuifion of that kinde, though in more familiar and
fcholafticall tearmes :   Namely that thefe bee the
two parts of Naturall Philofophie , the I N Q V I-
S I T I O N   O F   C A V S E S, and the P R O D V C T I O N
O F   E F F E C T S :   S P E C V L A T I V E,   and   O P E-
R A T I V E,   N A T V R A L L   S C I E N C E,   and   N A-
T V R A L L   P R V D E N C E.   For as in Ciuile matters
there is a wifedome of difcourfe, and a wifedome of
direction :   So is it in Naturall :   And heere I
will make a requeft, that for the latter (or at leaft for
a parte thereof) I may reuiue and reintegrate the
mifapplyed and abufed Name of   N A T V R A L L
M A G I C K E,   which in the true fenfe ,   is but
N A T V R A L L   W I S E D O M E,   or   N A T V R A L L
P R V D E N C E :   taken according to the ancient acception, purged from vanitie & fuperftition.   Now
although it bee true, and I know it well, that there
is an entercourfe betweene *Caufes* and *Effects* , fo as
both thefe knowledges *Speculatiue* & *Operatiue*, haue

a great

a great connexion betweene themfelues :   yet becaufe all true and frutefull NATVRALL PHI-
LOSOPHIE , hath *A double Scale* or *Ladder*, *Afcen-
dent* and *Defcendent*, afcending from experiments to
the *Inuention of caufes* ; and defcending from caufes,
to the *Inuention of newe experiments* ;   Therefore *I*
iudge it moft requifite that thefe two parts be feuerally confidered and handled.

        NATVRALL SCIENCE or THEORY
is deuided into PHISICKE and METAPHI-
SICKE, wherein *I* defire, it may bee conceiued,
that *I* vfe the word METAPHISICKE, in a differing
fenfe, from that, that is receyued: And in like manner *I* doubt not , but it will eafilie appeare to men
of iudgement , that in this and other particulers,
wherefoeuer my Conception & Notion may differ from the Auncient, yet *I* am ftudious to keepe
the Auncient Termes.   For hoping well to deliuer my felfe from miftaking , by the order and per-
fpicuous expreffing of that *I* doe propounde : *I* am
otherwife zealous and affectionate to recede as
little from Antiquitie, either in tearms or opinions,
as may ftand with truth, & the proficience of knowledge : And herein *I* cannot a little maruaile at the
Philofopher *Ariftotle* : that did proceede in fuch a
Spirit of difference & contradiction towards all Antiquitie, vndertaking not only to frame new wordes
of Science at pleafure : but to confound and extinguifh all ancient wifedome ; infomuch as hee neuer
                              nameth

nameth or mentioneth an Ancient Author or opinion, but to confute and reproue : wherein for glorie, and drawing followers and difciples, he tooke the right courfe, For certainly there commeth to paffe, & hath place in humane truth, that which was noted and pronounced in the higheft truth : *Veni in nomine Patris, nec recipitis Me, Si quis venerit in nomine suo, eum recipietis.* But in this diuine Aphorifme (confidering, to whom it was applied, Namely to *Antichrist,* the higheft deceiuer,) wee may difcerne well , that *the comming in a Mans owne name,* without regard of *Antiquitie,* or *paternitie* ; is no good figne of truth; although it bee ioyned with the fortune and fucceffe of an *Eum recipietis.* But for this excellent perfon *Ariftotle,* I will thinke of him , that hee learned that humour of his Scholler; with whom, it feemeth , hee did emulate, the one to conquer all Opinions, as the other to conquer all Nations. Wherein neuerthelefle it may bee, hee may at fome mens hands, that are of a bitter difpofition, get a like title as his Scholler did.

> *Fœlix terrarum Prædo, non vtile mundo*
> *Editus exemplum &c.* So
> *Fœlix doctrinæ Prædo.*

But to me on the other fide that do defire as much as lyeth in my Penne, to ground a fociable entercourfe betweene Antiquitie and Proficience, it feemeth beft, to keepe way with Antiquitie *vsque ad aras*; And therefore to retaine the ancient tearmes,

G g though

though I fometimes alter the vfes and definitions, according to the Moderate proceeding in Ciuill gouernment ; where although there bee fome alteration, yet that holdeth which *Tacitus* wifely noteth, *Eadem Magiſtratuum vocabula*.

To returne therefore to the vfe and acception of the tearme METAPHISICKE, as I doe nowe vnderftand the word; It appeareth by that which hath bene alreadie faide, that I intend, PHI-LOSOPHIA PRIMA : SVMMARIE PHILOSOPHIE and METAPHISICK, which heretofore haue beene confounded as one, to bee two diftinct things. For the one I haue made as a Parent, or common Aunceftor to all knowledge; And the other I haue now brought in, as a Branch or defcendent of NATVRALL SCIENCE; It appeareth likewife that I haue affigned to SVMMARIE PHILOSO-PHIE the common principles and Axiomes which are promifcuous and indifferent to feuerall Sciences : I haue affigned vnto it likewife the inqui-rie *touching the operation of the Relatiue and aduentiue Characters of Effences*, as *Quantitie, Similitude, Di-uerfitie, Pofsibilitie*, and the reft : with this diftin-ction, and prouifion : that they bee handled as they haue efficacie in Nature, and not logically. It appeareth likewife that NATVRAL THE-OLOGIE which heretofore hath beene handled confufedly with METAPHISICKE, I haue
inclofed

inclofed and bounded by it felfe. It is therefore now a queftion, what is left remaining for META-PHISICKE: wherein I may without preiudice preferue thus much of the cóceit of Antiquitie; that PHISICKE fhould contemplate that which is inherent in Matter, & therefore tranfitorie, and ME-TAPHISICKE, that which is abftraĉted & fixed. And againe that PHISICKE fhoulde handle that which fuppofeth in Nature onely a be-ing and mouing, and METAPHISICKE fhould handle that which fuppofeth furder in Na-ture, a reafon, vnderftanding, and platforme. But the difference perfpicuoufly expreffed, is moft familiar and fenfible. For as wee deuided NATVRALL PHILOSOPHY in GENERALL into the EN-QVIRIE of CAVSES & PRODVCTIONS of EFFECTS: So that part which concerneth the ENQVIRIE of CAVSES, wee doe fubdiuide, according to the receiued and found diuifion of CAVSES; The one part which is PHISICKE enquireth and handleth the MATERIALL & EFFICIENT CAVSES, & the other which is METAPHISICKE handleth the FOR-MAL and FINAL CAVSES.

PHISICKE, (taking it according to the deri-uation, & not according to our Idiome, for MEDI-CINE) is fcituate in a middle tearme or diftance between NATVRALL HISTORY & ME-TAPHISICKE. For NATVRAL HISTORY defcribeth the *varietie of things*: PHISICKE

the

the CAVSES, but VARIABLE Or RESPE-
CTIVE CAVSES; and METAPHISICKE the
FIXED and CONSTANT CAVSES.

Limus vt hic durescit, & hæc vt Cara liquescit,
Vno eodemque igni.

Fire is the cause of induration , but respectiue to
clay: Fire is the cause of colliquatiõ, but respectiue to
Waxe. But fire is noe constant cause either of induration or colliquation : So then the Phisicall causes
are but the Efficient and the Matter. PHISICKE
hath three parts, whereof two respect Nature *Vnited* or *collected*, the third contéplateth Nature *diffused*
or *distributed*. Nature is collected either into one entyer *Totall*, or else into the same *Principles* or *Seedes*.
So as the first doctrine is TOVCHING the CON-
TEXTVRE or CONFIGVRATION of
THINGS, as *De Mundo, de vniuersitate Rerum*. The
seconde is the Doctrine CONCERNING
the PRINCIPLES or ORIGINALS of
THINGS ; The third is the doctrine CON-
CERNING all VARIETIE and PARTI-
CVLARITIE of THINGS; whether it be of
the differing substances, or their differing qualities
and Natures; whereof there needeth noe enumera-
tion; this part being but as a GLOS or PARA-
PHRASE that attendeth vpon the Text of NA-
TVRALL HISTORY. Of these three I cannot
report any as deficient, In what truth or perfection
they are handled, I make not now any Iudgement:
But

But they are parts of knowledge not deferted by the Labour of Man.

For METAPHISICKE, we haue affigned vnto it the inquirie of FORMALL and FI-NALL CAVSES which affignation, as to the former of them may feeme to bee Nugatorie and voide, becaufe of the recciued and inueterate Opinion, that the inquifition of Man, is not competent to finde out *effentiall formes*, or *true differences*; of which opinion we will take this hold: That the Inuentiō of Formes is of al other Parts of Knowledge the worthieft to bee fought, if it bee Poffible to bee Found. As for the poffibili ie, they are ill difcoue-rers, that thinke there is no land when they can fce nothing but Sea. But it is manifeft, that *Plato* in his opinion of *Ideas*, as one that had a wit of elcuation fcituate as vpon a Cliffe, did defcry, *that formes were the true obiect of knowledge*; but loft the reall fruite of his opinion by confidering of formes, as abfolutely abftracted from Matter, & not confined and deter-mined by Matter : and fo turning his opinion vpon *Theologie*, wherewithall his Naturall Philofophy is infected. But if any man fhall keepe a continuall watchfull and feuere eye vpon action, operation, and the vfe of knowledge, hee may aduife and take Notice, what are the *formes*, the difclo-fures whereof are fruitful and important to the State of Man. For as to the *formes* of fubftāces (Man one-ly except, of whom it is faid, *Formauit hominem de li-mo terra, & fpirauit in faciem eius fpiraculum vitæ*, and not as of all other creatures, *Producant aqua, produ-cat terra, the formes of Subſtances* I ſay (as they are nowe by compounding and tranſplanting multi-plied) are ſo perplexed, as they are not to bee en-quired. Noe more then it were either poſſible or to purpoſe, to ſeeke in groſſe *the formes of thoſe Soundes which make wordes*, which by compoſition and tranſpoſition of Letters are infinite. But on the otherſide, to enquire *the forme of thoſe Soundes or Voices which make ſimple Letters* is eaſily comprehēſi-ble, and being knowen, induceth and manifeſteth *the formes of all words*, which conſiſt, &are compoun-ded of them; in the ſame maner to enquire *the forme* of a Lyon, of an Oake, of Gold: Nay of Water, of Aire, is a vaine purſuite. But to enquire *the formes* of Sence, of voluntary Motion, of Vegetation, of Co-lours, of Grauitie and Leuitie, of Denſitie, of Tenui-tie, of Heate, of Cold, & al other Natures and quali-ties, which like an *Alphabet* are not many, &of which the eſſences (vpheld by Matter) of all creatures doe cōſiſt: To enquire *I* ſay *the true formes* of theſe, is that part of METAPHISICKE, which we now define of. Not but that PHISICKE doth make enquirie, and take conſideration of the ſame Natures, but how? Onely, as to the *material and efficient cauſes* of them, and not as to *the formes*. For example, if the *cauſe* of *whiteneſſe* in *Snowe* or *froth* be enquired, and it be rendred thus: *That the ſubtile intermixture of Ayre and water is the cauſe*, it is well rendred, but ne-uerthe-

uertheleſſe is this *the forme* of *whiteneſſe ?* Noe but it is *the efficient ,* which is euer but *vehiculum formæ.* This, part of METAPHISICKE : I doe not finde laboured and performed, whereat I maruaile not , Becauſe I hold it not poſſible to bee inuented by that courſe of inuention which hath beene vſed, in regard that men   (which is the Roote of all error) haue made too untimely a departure, and to remote a receſſe from particulars.

Metaphi-
ſica ſiue
De formis
& Fini-
bus Rerū.

But the vſe of this part of M E T A P H I S I C K E which *I* report as deficient, is of the reſt the moſt excellent in two reſpects: The one becauſe it is the dutie and vertue of all knowledge to abridge the infinitie of indiuiduall experience , as much as the conception of truth will permit , and to remedie the complaint of *vita breuis, ars longa ;* which is performed by vniting the Notions and conceptions of Sciences: For knowledges are as P Y R A M I D E S , whereof H I S T O R Y is the B A S I S : So of N A T V- R A L P H I L O S O P H Y the B A S I S is N A T V R A L H I S T O R Y: The S T A G E next the B A S I S is P H I- S I C K E : The STAGE next the V E R T I- C A L P O I N T is METAPHISICKE: As for the V E R T I C A L L P O I N T, *Opus quod o-* *peratur deus a principio vſque ad finem, the Summary law* *of Nature ,* wee knowe not whether Mans enquirie can attaine vnto it. But theſe three be the true *Stages* of knowledge, and are to them that are depraued no better then the Gyants Hilles.

Ter

Ter funt conati imponere Pelio Offam:
Scilicet atque Offæ frondofum involuere Olympum.

But to thofe which referre all thinges to the
Glorie of G O D , they are as the three acclama-
tions : *Sancte, Sancte, Sancte*: holy in the defcription or dilatation of his workes, holy in the connexion, or concatenation of them, and holy in the
vnion of them in a perpetuall and vniforme lawe.
And therefore the fpeculation was excellent in *Par-*
menides and *Plato* , although but a fpeculation in
them, That all things by fcale did afcend to vnitie.
So then alwaies that knowledge is worthieft, which
is charged with leaft multiplicitie, which appeareth
to be M E T A P H I S I C K E, as that which confidereth *the fimple formes or differēces of things*, which
are few in number, and *the degrees* and *coordinations*
whereof, make all this varietie : The fecond refpect
which valueth and commendeth this part of M E-
T A P H I S I C K E is, that it doth enfranchife the
power of Man vnto the greateft libertie, and poffi-
bilitie of workes and effects. For Phificke carri-
eth men in narrow and reftrained waies, fubiect to
many accidents of impedimēts, imitating the ordinarie flexuous courfes of Nature, But *Latæ vndique*
funt fapientibus viæ: To fapience (which was anciently defined to be *Rerum diuinarum, & humanarum fci-*
entia) there is euer choife of Meanes. For *Phificall*
caufes

caufes giue light to newe inuention in *Simili materia*;
But whofoeuer knoweth any *forme* knoweth the
vtmoft *pofsibilitie* of *fuperinducing* that *Nature* vpon
any varietie of Matter, and fo is leffe reftrained in o-
peration, either to the *Bafis* of the *Matter*, or the *condition* of the *efficient* : which kinde of knowledge
Salomon likewife, though in a more diuine fort elegantly defcribeth, *Non arctabuntur greffus tui, & cur-
rens non habebis offendiculum*. The waies of fapience
are not much lyable , either to particularitie or
chance.

The 2. part of METAPHISICKE is the ENQI-
RY of FINAL CAVSES, which I am moued to
report, not as omitted, but as mifplaced; And yet if
it were but a fault in order, I would not fpeake of it.
For order is matter of illuftration , but pertaineth
not to the fubftance of Sciences *:* But this mifpla-
cing hath caufed a deficience, or at leaft a great im-
proficience in the Sciences themfelues. For the
handling of *finall caufes* mixed with the reft in *Phificall enquiries*, hath intercepted the feuere and diligent enquirie of all *reall and phificall caufes*, and giuen men the occafion , to ftay vpon thefe *fatiffactorie and fpecious caufes*, to the great arreft and preiudice of furder difcouerie . For this I finde done
not onely by *Plato* , who euer ancreth vppon
that fhoare , but by *Ariftotle*, *Galen* , and others,
which doe vfually likewife fall vppon thefe flatts *of
difcourfing caufes* ; For to fay that the haires of the
H h Eye-

Eye-liddes are for a quic-sette and fence about the
Sight : Or, That the firmeneſſe of the Skinnes and Hides
of liuing creatures is to defend them from the extremities of heate or cold : Or , That the bones are for
the columnes or beames , whereupon the Frame of the
bodies of liuing creatures are built ; Or , That the
leaues of trees are for protecting of the Fruite; Or,
That the cloudes are for watering of the Earth ; Or,
That the ſolidneſſe of the Earth is for the ſtation and
Manſion of liuing creatures : and the like, is well inquired & collected in METAPHISICKE, but
in PHISICKE they are impertinent. Nay, they
are indeed but *Remoraes* and hinderances to ſtay and
ſlugge the Shippe from furder ſayling , and haue
brought this to paſſe, that the ſearch of the *Phiſicall
Cauſes* hath beene neglected , and paſſed in ſilence.
And therefore the Natural Philoſophie of *Democritus*, and ſome others who did not ſuppoſe a *Minde* or
Reaſon in the frame of things, but attributed the *form
thereof able to maintaine it ſelf to infinite eſſaies or proofes
of Nature*, which they tearme *fortune* ; ſeemeth to
mee (as farre as I can iudge by the recitall and
fragments which remaine vnto vs) in particularities
of Phiſicall cauſes more reall and better enquired
then that of *Ariſtotle* and *Plato*, whereof both intermingled *final cauſes*, the one as a part of *Thelogie*, and
the other as a part of *Logicke*, which were the *fauou-
rite ſtudies* reſpectiuely of both thoſe perſons. Not
becauſe

becaufe thofe *finall caufes* are not true, and worthy to bee inquired, beeing kept within their owne prouince; but becaufe their excurfions into the limits of *Phificall caufes*, hath bred a vaftneffe and folitude in that tract. For otherwife keeping their precincts and borders, men are extreamely deceiued if they thinke there is an Enmitie or repugnancie at all betweene them: For the caufe rendred that *the haires about the Eye-liddes are for the fafegard of the fight*, doth not impugne the caufe rendred, *that Pilofitie is incident to Orifices of Moifture : Mufcofi fontes &c*. Nor the caufe rendred *that the firmeneffe of hides is for the armour of the body againft extremities of heate or cold* : doth not impugne the caufe rendred , *that contraction of pores is incident to the outwardeft parts; in regard of their adiacence to forreine or vnlike bodies*, and fo of the reft; both caufes beeing true and compatible, the one declaring an *intention*, the other a *confequence* onely. Neither doth this call in queftion or derogate from diuine Prouidence, but highly confirme and exalt it. For as in ciuill actions he is the greater and deeper pollitique, that can make other men the Inftruments of his will and endes, and yet neuer acquaint them with his purpofe : So as they fhall doe it, and yet not knowe what they doe, then hee that imparteth his meaning to thofe he employeth: So is the wifdome of God more admirable, when

H h 2   Nature

Nature intendeth one thing, and Prouidéce draweth forth another; then if hee had communicated
to particular Creatures and Motions the Chara-
&ers and Impreffions of his Prouidence; And thus
much for M E T A P H I S I C K E,the later part wherof,I allow as extant,but wifh it confined to his proper place.

Neuertheleffe there remaineth yet another
part of N A T V R A L L P H I L O S O P H I E,
which is commonly made a principall part , and
holdeth ranke with P H I S I C K E fpeciall and
M E T A P H I S I C K E : which is *Mathematicke,*
but *I* think it more agreable to the Nature of things,
and to the light of order, to place it as a Branch of
Metaphificke : For the fubie&t of it being *Quantitie,*
not *Quantitie Indefinite:* which is but a *Relatiue,* and
belongeth to *Philofophia Prima*(as hath beene faid,)
but *Quantitie determined, or proportionable*, it appeareth to bee one of the *effentiall formes* of things;
as that, that is caufatiue in Nature of a number
of Effe&ts, infomuch as wee fee in the Schooles
both of *Democritus,* and of *Pithagoras,* that the
one *did afcribe Figure to the firft feedes of things,* and
the other *did fuppofe numbers to bee the principalles
and originalls of things* ; And it is true alfo that of
all other formes (as wee vnderftand formes) it is
the moft abftra&ted, and feparable from matter
and therefore moft proper to *Metaphificke*; which
hath

hath likewife beene the caufe, why it hath beene better laboured, and enquired, then any of the other *formes*, which are more immerfed into Matter. For it beeing the Nature of the Minde of Man (to the extreame preiudice of knowledge) to delight in the fpacious libertie of generalities, as in a champion Region; and not in the inclofures of particularitie; the MATHEMATICKS of all other knowledge were the goodlieft fieldes to fatiffie that appetite. But for the placing of this Science, it is not much Materiall: onely we haue endeuoured in thefe our Partitions to obferue a kind of perfpectiue, that one part may caft light vpon another.

The MATHEMATICKS are either PVRE, or MIXT: To the PVRE MATHEMATICKS are thofe Scieces belonging, which handle *Quantitie determinate* meerely feuered from any Axiomes of NATVRALL PHILOSOPHY: and thefe are two, GEOMETRY and ARITHMETICKE, The one handling Qnantitie continued, and the other diffeuered. MIXT hath for fubiect fome Axiomes or parts of Naturall Philofopie: and confidereth Quantitie determined, as it is auxiliarie and incident vnto them. For many parts of Nature can neither be inuented with fufficient fubtiltie, nor demonftrated with fufficient perfpicuitie, nor accommodated

vnto vfe with fufficient dexteritie, without the aide
and interueyning of the Mathematicks : of which
forte are *Perfpecîiue*, *Muficke*, *Aftronomie*, *Cofmogra-*
phie, *Archîteéîure*, *Inginarie*, and diuers others. In the
Mathematicks, I can report noe deficience, except it
be that men doe not fufficiently vnderftand the excellent vfe of *the pure Mathematicks*, in that they doe
remedie and cure many defeéts in the Wit, and
Facultics Intelleéîuall. For, if the wit bee to
dull, they fharpen it : if to wandring, they fix
it : if to inherent in the fenfe, they abftraét it.
So that, as Tennis is a game of noe vfe in it felfe,
but of great vfe, in refpeét it maketh a quicke
Eye, and a bodie readie to put it felfe into all
Poftures : So in the Mathematickes, that vfe which
is collaterall and interuenient, is no leffe worthy,
then that which is principall and intended. And
as for the *Mixt Mathematikes* I may onely make this
prediéîion, that there cannot faile to bee more
kindes of them, as Nature growes furder difclofed.
Thus much of N A T V R A L S C I E N C E, or
the part of Nature S P E C V L A T I V E.

For N A T V R A L L P R V D E N C E, or the
part O P E R A T I V E of N A T V R A L L P H I-
L O S O P H Y, we will deuide it into three parts,
E X P E R I M E N T A L, P H I L O S O P H I C A L
and M A G I C A L, which three parts A C T I V E
haue a correfpondéce and Analogie with the three
parts S P E C V L A T I V E: N A T V R A L H I-
S T O R Y,

STORY, PHISICKE, and METAPHI-
SICKE: For many operations haue bin inuented
sometime by a casuall incidence and occurrence,
sometimes by a purposed experiment: and of those
which haue bene found by an intentionall experi-
mēt, some haue bin found out by varying or extending the same experiment, some by transferring and
compounding diuers experiments the one into the
other, which kind of inuention an Emperique may
manage. Againe by the knowledge of Phisicall causes, there cannot faile to followe, many indications
and designations of new particulers, if men in their
speculation will keepe one eye vpon vse & practise.
But these are but Coastings along the shoare, *Pre-*
mendo littus iniquum, For it seemeth to me, there can
hardly bee discouered any radicall or fundamentall
alterations, and innouations in Nature, either by
the fortune & essayes of experiments, or by the light
and direction of Phisical causes. If therfore we haue
reported METAPHISICKE deficient, it must followe, that wee doe the like of NATVRAL MA-
GICKE, which hath relation thereunto. For as for
the NATVRAL MAGICKE whereof now there
is mention in books, containing certaine credulous and superstitious conceits and obseruations
of *sympathies,* and *Antipathies* and *hidden Pro-*
prieties, and some friuolous experiments, strange
rather by disguisement, then in themselues, It is as
fardiffering in truth of Nature, from such a know-
edge as we require, as the storie of King *Arthur*
of

Naturalis
Magiasiue
Phisica O-
perativa
Maior.

of *Brittaine*, or *Hughe* of *Burdeaux*, differs from
Cæfars commentaries in truth of ftorie. For it is
manifeft that *Cæfar* did greater things *de vero*, then
thofe *Imaginarie Heroes* were fained to doe. But
hee did them not in that fabulous manner. Of this
kinde of learning the fable of Ixion was a figure:
who defigned to enioy *Iuno* the Goddeffe of power: and in ftead of her, had copulation with a Cloud:
of which mixture were begotten Centaures, and
Chymeraes. So whofoeuer fhall entertaine high
and vapourous imaginations, in fteede of a laborious and fober enquirie of truth fhall beget hopes
and Beliefes of ftrange and impoffible fhapes. And
therefore wee may note in thefe Sciences, which
holde fo much of imagination and Beliefe, as this
degenerate Naturall Magicke, Alchimie, Aftrologie, and the like, that in their propofitions, the defcription of the meanes, is euermore monftrous,
then the pretence or ende . For it is a thing more
probable, that he that knoweth well the Natures of
Waight, *of Colour*, *of Pliant*, and *fragile* in refpect of
the hammer, of *volatile* and *fixed* in refpect of the
fire, and the reft, may fuperinduce vpon fome Mettall the Nature, and forme of Gold by fuch *Me-*
chanique as longeth to the production of the Naturs
afore rehearfed , then that fome graynes of the
Medecine proiected, fhould in a fewe Moments
of time, turne a Sea of Quick filuer or other Materiall

teriall into Gold. So it is more probable that he that
knoweth the Nature of *Arefaction*; the Nature of
aſsimilation, of nouriſhment to the thing nouriſhed;
the Maner of *encreaſe, and clearing of ſpirits* : the Maner of the *depredations* , *which Spirits make vpon the
humours and ſolide parts* : ſhall, by Ambages of diets,
bathings, annointings, Medecines, motions, and
the like , prolong life , or reſtore ſome degree of
youth or viuacitie, then that it can be done with the
vſe of a fewe drops, or ſcruples of a liquor or receite.
To conclude therefore, the true N A T V R A L L
M A G I C K E, which is that great libertie and Latitude of operation, which dependeth vppon the
knowledge of *formes*, I may report deficient, as the
Relatiue thereof is; To which part if we be ſerious,
and incline not to vanities and plauſible diſcourſe,
beſides the deriuing and deducing the operations
themſelues from M E T A P H I S I C K E, there are
pertinent two points of much purpoſe, the one by
way of preparation , the other by way of caution:
The firſt is, that there be made a *Kalender reſembling
an Inuentorie of* the eſtate of man, containing all the   *Inuentari-*
inuentions, (being the works or fruits of Nature or   *um Opum*
Art) which are now extant, and whereof man is al-   *humana-*
readie poſſeſſed, out of which doth naturally reſult   *rum.*
a Note , what things are yet held impoſſible , or
not inuented, which *Kalender* will bee the more
artificiall and ſeruiceable, if to euery *reputed impoſſibilitie*, you adde what thing is extant , which
I i   commeth

commeth the neareſt in degree to that *Impoſſibili-tie*;to the ende, that by theſe *Optatiues* and *Potenti-alls*, Mans enquirie may bee the more awake in di-ducing directiõ of works from the ſpeculatiõ of cau-ſes. And ſecondly that thoſe *experimẽts* be not one-ly eſteemed which haue an immediate & preſẽt vſe, but thoſe principally which are of moſt vniuerſall conſequence for inuention of other experimẽts, & thoſe which giue moſt light to the Inuẽtion of cau-ſes; for the Inuẽtion of the Mariners Needle, which giueth the direction, is of noe leſſe beneſit for Naui-gation, then the inuention of the ſailes which giue the Motion.

Thus haue I paſſed through NATVRALL PHILOSOPHIE, and the deficiences there-of; wherein if I haue differed from the ancient, and receiued doctrines, and thereby ſhall moue contra-diction; for my part, as I affect not to diſſent, ſo I purpoſe not to contend; If it be truth.

- - *Non canimus ſurdis reſpondent omnia ſyluæ;*
The voice of Nature will conſent, whether the voice of Man doe or noe. And as *Alexander Bergia* was wont to ſay of the expedition of the french for *Naples*, that they came with Chaulke in their hands to marke vp their lodgings, and not with weapons to fight : So I like better that entrie of truth which commeth peaceably with Chaulke, to marke vp thoſe Mindes, which are capable to lodge and har-bour it, then that which commeth with pugnaci-tie and contention.

But there remaineth a diuiſion of Naturall Phi-loſophy

lofophy according to the *Report of the Enquirie*, and nothing concerning the Matter or fubiect, and that is POSITIVE and CONSIDERA-TIVE: when the enquirie reporteth either an *Affertion*, or a *Doubt*. Thefe *doubts* or *Non Liquets*, are of two forts, *Particular* and *Totall*. For the firft wee fee a good example thereof in *Ariftotles* Problemes, which deferued to haue had a better continuance, but fo neuertheleffe, as there is one point, whereof warning is to be giuen and taken; The Regiftring of doubts hath two excellent vfes : The one that it faueth Philofophy from Errors & falfhoods: when that which is not fully appearing, is not collected into affertion, whereby Error might drawe Error, but referued in doubt. The other that the entrie of doubts are as fo many fuckers or fponges, to drawe vfe of knowledge, infomuch as that which if doubts had not preceded, a man fhould neuer haue aduifed, but paffed it ouer without Note, by the fuggeftion and follicitation of doubts is made to be attended and applied. But both thefe commodities doe fcarcely counteruaile an *Inconuenience*, which wil intrude it felfe if it be not debarred, which is that when a doubt is once receiued, men labour rather howe to keepe it a doubt ftill, then howe to folue it, and accordingly bend their wits. Of this we fee the familiar example in Lawyers and Schollers, both which if they haue once admitted a doubt, it goeth euer after Authorized for a doubt. But that vfe of wit and knowledge is to be

allowed

allowed which laboureth to make doubtfull thinges certaine, and not thofe which labour to make certaine things doubtfull. Therefore thefe *Kalenders of doubts*, I commend as excellent things, fo that there be this caution vfed, that when they bee throughly fifted & brought to refolution, they bee from thence forth omitted, decarded, and not continued to cherifh and encourage men in doubting. To which *Kalender* of doubts or problemes, I aduife be annexed another *Kalender* as much or more Materiall, which is a *Kalender of popular Errors*, I meane chiefly, in naturall Hiftorie fuch as paffe in fpeech & conceit, and are neuertheleffe apparantly detected & couicted of vntruth, that Mans knowledge be not weakened nor imbafed by fuch droffe and vanitie. As for the *Doubts or Non liquets generall or in Totall*, I vnderftand thofe differences of opinions touching the principles of Nature, and the fundamentall points of the fame, which haue caufed the diuerfitie of Sects, Schooles, and Philofophies, as that of *Empedocles, Pythagoras, Democritus, Parmenides*, and the reft. For although *Ariftotle* as though he had bin of the race of the *Ottomans*, thought hee could not raigne, except the firft thing he did he killed all his Brethren; yet to thofe that fecke *truth* and not *Magiftralitie*, it cannot but feeme a Matter of great profit, to fee before them the feueral opinions touching the foundations of Nature, not for any exact truth that can be expected in thofe Theories : For as the
same

*Continua-
tio Proble-
matum in
Natura.
Catalogus
Falfitatū
graffantiū
in hiftoria
Naturæ.*

fame *Phenomena* in Aftronomie are fatiffied by the
receiued Aftronomie of the diurnall Motion , and
the proper Motions of the Planets, with their *Eccen-*
triques and *Epicicles* and likwife by the Theorie of
Copernicus, who fuppofed the Earth to moue; & the
Calculations are indifferently agreeable to both: So
the ordinarie face and viewe of experience is many
times fatiffied by feuerall Theories & Philofophies,
whereas to finde the reall truth requireth another
manner of feueritie & attention . For, as *Ariftotle*
faith that children at the firft will call euery woman
mother : but afterward they come to diftinguifh
according to truth: So Experience, if it be in child-
hood, will call *euery Philofophie Mother*; but when it
commeth to ripeneffe, it will difcerne the true Mother. So as in the meane time it is good to fee the
Seuerall Gloffes and Opinions vpon Nature, wher- *De Anti-*
of it may bee euery one in fome one point , hath *quis Phi-*
feene clearer then his fellows; Therfore I wifh fome *lofophijs.*
collection to be made painfully and vnderftanding-
ly *de Antiquis Philofophijs* out of all the poffible light
which remaineth to vs of them . Which kinde of
worke I finde deficient. But heere I muft giue
warning , that it bee done diftinctly and feuere-
ly ; The Philofophies of euery one throughout by
themfelues; and not by titles packed, and fagotted
vp together, as hath beene done by *Plutarch.* For
it is the harmonie of a Philofophie in it felfe, which
giueth it light and credence ; whereas if it bee fin-

gled and broken, it will feeme more forraine and diffonant. For as, when I read in *Tacitus*, the Actions of *Nero*, or *Claudius*, with circumftances of times, inducements and occafions, I finde them not fo ftrange: but when I reade them in *Suetonius Tranquillus* gathered into tytles and bundles, and not in order of time, they feeme more monftrous and incredible; So is it of any Philofophy reported entier, and difmembred by Articles. Neither doe I exclude opinions of latter times to bee likewife reprefented, in this Kalender of Sects of Philofophie, as that of *Theophraftus Paracelfus*, eloquently reduced into an harmonie, by the Penne of *Seuerinus* the *Dane*: And that of *Tylefius*, and his Scholler *Donius*, beeing as a Paftorall Philofophy, full of fenfe, but of no great depth. And that of *Fracaftorius*, who though hee pretended not to make any newe Philofophy, yet did vfe the abfolutenefle of his owne fenfe, vpon the olde. And that of *Gilbertus*, our countreyman, who reuiued, with fome alterations, and demonftrations, the opinions of *Xenophanes*, and any other worthy to be admitted.

Thus haue we now dealt w̃ two of the three *beames* of Mans knowledge, that is *Radius Directus*, which is referred to Nature, *Radius Refractus*, which is referred to God, and cannot report truely becaufe of the inequalitie of the *Medium*. There refteth *Radius Reflexus*, whereby Man beholdeth and contemplateth himfelfe.

WE come therefore now to that knowledge, whereunto the ancient Oracle directeth vs, which

which is, *the knowledge of our selues:* which deserueth the more accurate handling, by howe much it toucheth vs more neerely . This knowledge as it is the end and Terme of Naturall Philosophy *in the intention of Man :* So notwithstanding it is but a portion of Naturall Philosophy *in the continent of Nature:* And generally let this be a Rule, that all partitions of knowledges,be accepted rather for *lines & veines,* then for *sections and separations:* and that the continuance and entirenes of knowledge be preserued. For the contrary hereof hath made particular Sciences, to become barren, shallow, & erronious: while they haue not bin Nourished and Maintained from the cōmon fountaine: Sowe see *Cicero* the Orator complained of *Socrates* and his Schoole, that he was the first that separated Philosophy , and Rhetoricke, whereupon Rhetorick became an emptie & verball Art. So wee may see that the opinion of *Copernicus* touching the rotation of the earth, which Astronomie it self cānot correct, because it is not repugnant to any of the *Phainomena,* yet Naturall Philosophy may correct. So we see also that the Science of *Medicine,*if it be destituted & forsaken by *Natural Philosophy* , it is not much better then an Empeirical practize: with this reseruation therefore we proceed to HVMANE PHILOSOPHY or HVMANITIE, which hath two parts : The one considereth Man *segregate,or distributiuely:* The other *congregate* or *in societie.* So as HVMANE PHILOSOPHY is either SIMPLE and PARTICVLAR,

or coniugate and Ciuile; Hvmanitie Particvlar confifteth of the fame parts, whereof Man
confifteth, that is, of Knovvledges Which
Respect The Body, & of Knovvledges that respect the mind. But before
we *diftribute* fo far, it is good to *conftitute*. For I doe
take the confideration in generall , and at large of
hvmane natvre to be fit to be emancipate, &
made a knowledge by it felf; Not fo much in regard
of thofe delightfull and elegant difcourfes , which
haue bin made of the dignitie of Man , of his mife-
ries , of his ftate and life , and the like *Adiuncts
of his common and vndeuided Nature* , but chiefely in regard of the knowledge concerning the
sympathies and concordances be-
tvveene The mind and body , which
being mixed , cannot be properly affigned to the
fciences of either.

    This knowledge hath two branches; for as all
leagues and Amities confift of mutuall *Intelligence*,
and mutuall *Offices* So this league of mind and body, hath thefe two parts, *How the one d fclofeth the o-
ther* , *and how the one worketh vpon the other* , *Difcouerie*, & *Impreffion*. The former of thefe hath be-
gotté two Arts, both of *Predictió* or *Prenotion* where
of the one is honoured with the enquirie of *Ariftotle*, & the other of *Hippocrates* . And although they
haue of later time beene vfed to be coupled with
                      fuperftitious

superstitious and fantasticall arts ; yet being purged
and restored to their true state ; they haue both of
them a solide ground in nature, and a profitable vse
in life. The first is PHYSIOGNOMIE, which difcouereth the disposition of the mind, by the Lyneaments of the bodie. The second is the EXPOSI-
TION OF NATVRALL DREAMES, which
difcouereth the state of the bodie, by the imaginations of the minde. In the former of thefe, I note a
deficience. For *Ariftotle* hath verie ingenioufly,
and diligently handled the factures of the bodie, but
not the gestures of the bodie; which are no leffe
comprehenfible by art, and of greater vfe, and aduantage. For the Lyneaments of the bodie doe difclofe the difposition and inclination of the minde in
generall; but the Motions of the countenance and
parts, doe not onely so, but doe further difclofe the
prefent humour and state of the mind & will. For as
your Maieftie fayth most aptly and elegantly; *As
the Tongue speaketh to the Eare, so the gesture speaketh to
the Eye.* And therefore a number of fubtile perfons,
whofe eyes doe dwell vpon the faces and fafhions
of men; doe well know the aduantage of this obferuation; as being most part of their abilitie; neither can it bee denied, but that it is a great difcouerie of difsimulations, and a great direction in Bufineffe.

The later Braunch, touching IMPRESSION
hath not beene collected into Art ; but hath beene
handled difperfedly; and it hath the fame relation

or *Antistrophe*, that the former hath. For the con-
sideration is double, EITHER HOVV, AND
HOVV FARRE THE HVMOVRS AND AF-
FCTS OF THE BODIE, DOE ALTER OR
WORKE VPON THE MIND; or againe, HOVV
AND HOVV FARRE THE PASSIONS, OR
APPREHENSIONS OF THE MINDE, DOE
ALTER OR WORKE VPON THE BODIE.
The former of thefe, hath beene enquired and confidered, as a part, and appendix of Medicine, but
much more as a part of Religion or fuperftition.
For the Fhifitian prefcribeth Cures of the minde in
Phrenfies, and melancholy pafsions; and pretendeth
alfo to exhibite Medicines to exhilarate the minde,
to confirme the courage, to clarifie the wits, to corroborate the memorie, and the like; but the fcruples
and fuperftitions of Diet, and other Regiment of
the body in the fect of the *Pythagoreans*, in the Herefy
of the *Manicheas*, and in the Lawe of *Mahumet* doe
exceede; So likewife the ordinances in the Ceremoniall Lawe, interdicting the eating of the blood,
and the fatte; diftinguifhing between beafts cleane
and vncleane for meat; are many and ftrict. Nay,
the faith it felfe, being cleere and ferene from all
cloudes of Ceremonie, yet retaineth the vfe of fa-
ftings, abftinences, and other Macerations and hu-
miliations of the bodie, as things reall, & not figura-
tiue. The roote and life of all which prefcripts, is
(befides the Ceremonie,) the confideration of that
dependancie, which the affections of the mind are
                                          fubmitted

submitted vnto, vpon the state and disposition of the bodie. And if any man of weake iudgement doe conceiue, that this suffering of the minde from the bodie, doth either question the Immortalitie, or derogate from the soueraigntie of the soule : hee may be taught in easie instances, that the Infant in the mothers wombe, is compatible with the mother, and yet separable : And the most absolute Monarch is sometimes ledde by his seruants, and yet without subiection As for the reciprocall knowledge, which is the operation of the conceits and passions of the minde vppon the bodie ; We see all wise Phisitians in the prescriptions of their regiments to their Patients, doe euer consider *Accidentia animi* : as of great force to further or hinder remedies, or recoueries ; and more specially it is an inquirie of great depth and worth, concerning I M A G I N A T I O N, how, and howe farre it altereth the bodie proper of the Imaginant. For although it hath a manifest power to hurt, it followeth not, it hath the same degree of power to helpe. No more than a man can conclude, that because there be pestilent Ayres, able sodainely to kill a man in health ; therefore there should bee soueraigne ayres, able sodainly to cure a man in sicknesse. But the inquisition of this part is of great vse, though it needeth, as *Socrates* sayd, *A Delian diuer,* being difficult & profound. But vnto all this knowledge D E C O M M V N I V I N C V L O, of the Concordances betweene the Mind and the bodie : that part of Enquirie is most necessarie, which conside-

reth

reth of the *Seates*, and *Domiciles* which the feuerall faculties of the minde, doe take and occupate in the Organs of the bodie, which knowledge hath been attempted, and is controuerted, and deferueth to bee much better inquired. For the opinion of *Plato*, who placed *the Vnderſtanding in the Braine* ; *Animoſitie*, (which hee did vnfitly call *Anger*, hauing a greater mixture with *Pride*) *in the Heart* ; and *Concupiſcence* or *Senſualitie in the Liuer*, deferueth not to bee defpifed, but much leſſe to be allowed. So then we haue conſtituted (as in our own wiſh and aduiſe) the inquirie TOVCH-ING HVMANE NATVRE ENTYER ; as a iuſt portion of knowledge, to be handled apart.

The knowledge that concerneth mans bodie, is diuided as the good of mans bodie is diuided, vnto which it referreth. The good of mans body, is of foure kindes ; *Health*, *Beautie*, *Strength*, and *Pleaſure*. So, the knowledges are *Medicine*, *or Art of Cure* : *Art of Decoration* ; which is called *Coſme-rike* : *Art of Aƈiuitie*, which is called *Athletike* : and *Art Voluptuarie*, which *Tacitus* truely calleth *Eruditus Luxus*. This Subieƈt of mans bodie, is of all other thinges in Nature, moſt fufceptible of remedie : but then that Remedie is moſt fufceptible of errour. For the fame *Subtilitie* of the fubieƈt, doth caufe large poſsibilitie, and eaſie fayling : and therefore the enquirie ought to be the more exaƈt.

To fpeak therfore of *Medicine*, & to refume that we haue fayd, afcending a litle higher ; The ancient opinion

nion that *Man* was *Microcofmus,* an Abftra&t or Modell of the world, hath beene fantaftically ftreyned
by *Paracelfus,* and the Alchimifts, as if there were to
be found in *mans body* certaine correfpondences, &
parallells, which fhold haue refpe&t to all varieties of
things, as ftarres, planets, minerals, which are extant
in the great world. But thus much is euidently true,
that of all fubftances, which Nature hath produced,
mans bodie is the moft extreamly compounded. For
we fee hearbs & plants are norifhed by earth & water; Beafts for the moft part, by hearbs & fruits; Man
by the flefh of Beafts, Birds, Fifhes, Hearbs, Grains,
Fruits, Water, & the manifold alterations, drefsings,
and preparations of thefe feuerall bodies, before
they come to be his food & aliment. Adde hereunto
that Beafts haue a more fimple order of life, and lefle
change of Affe&tions to worke vppon their bodies, whereas man in his *Manfion,* fleepe, exercife,
pafsions, hath infinit variations; and it cannot be denied, but that the *bodie of Man* of all other things, is of
the moft compounded Maffe. *The foule* on the other
fide is the fimpleft of fubftances, as is well expreffed.

Purumq; reliquit
Æthereum fenfum, atque Aurai fimplicis ignem.

So that it is no maruaile, though *the foule* fo placed,
enioy no reft, if that principle be true, that *Motus*
rerum eft rapidus extra locum, Placidus in loco. But
to the purpofe, this variable compofition of mans
bodie hath made it as an Inftrument eafie to
diftemper; and therefore the Poets did well to
K k 3 conioyne

conioyne MVSICKE and MEDICINE in *Apollo*, becaufe the Office of Medicine, is but to tune this curious Harpe of mans bodie, and to reduce it to Harmonie. So then the *Subiect* being fo *Variable*, hath made the *Art* by confequent more *coniecturall*, and the Art being Coniecturall, hath made fo much the more place to bee left for impofture. For almoft all other Arts and Sciences, iudged by Acts, or Mafter peeces, as I may terme them, and not by the fucceffes, and euents. The Lawyer is iudged by the vertue of his plea-ding, and not by the yffue of the caufe: The Mafter in the Shippe, is iudged by the directing his courfe aright, and not by the fortune of the Voyage: But the Phifitian, and perhaps the Politique, hath no particular Acts demonftratiue of his abilitie, but is iudged moft by the euent : which is euer but as it is taken ; for who can tell if a Patient die or recouer, or if a State be preferued, or ruyned, whether it be Art or Accident? And therefore many times the Im-poftor is prized, and the man of vertue taxed. Nay, we fee weakeneffe and credulitie of men, is fuch, as they will often preferre a Montabanke or Witch, before a learned Phifitian. And therefore the Poets were cleere fighted in difcerning this ex-treame folly, when they made *Æfculapius*, and *Circe*, Brother and Sifter, both Children of the Sunne, as in the verfes.

Ipfe repertorem medicinæ talis & artis,

Fulmine Phœbigenam *ftygias detrufit ad vndas,*

And

And againe.

Diues inacceßos vbi Solis filia *Lucos, &c.*

For in all times in the opinion of the multitude, Witches, and old women, and Impoſtors haue had a Competicion with Phiſitians. And what followeth? Euen this that Phiſitians ſay to themſelues, as *Salomon* expreſſeth it vpon an higher occaſion: *If it befall to me, as befalleth to the fooles, why ſhould I labour to be more wiſe?* And therefore I cannot much blame Phiſitians, that they vſe commonly to intend ſome other Art or practiſe, which they fancie, more than their profeſsion. For you ſhall haue of them: Antiquaries, Poets, Humaniſts, Stateſ-men, Marchants, Diuines, and in euerie of theſe better ſeene, than in their profeſsion, & no doubt, vpon this ground that they find, that mediocrity & excellency in their Art, maketh no difference in profite or reputation towards their fortune: for the weakeneſſe of Patients, and ſweetneſſe of life, and Nature of hope maketh men depend vpon Phiſitians, with all their defects. But neuertheleſſe, theſe things which we haue ſpoken of, are courſes begotten betweene a little occaſion, and a great deale of ſloath and default: for if we will excite and awake our obſeruation, we ſhall ſee in familiar inſtances, what a predominant facultie, *The Subtiltie* of *Spirite*, hath ouer the *Varietie* of *Matter*, or *Fourme*: Nothing more variable then faces and countenances: yet men can beare in memorie the infinite diſtinctions of them. Nay, a Painter with a fewe ſhelles of

Kk 4   colors

colours, and the benefite of his Eye, and habite of his imagination can imitate them all that euer haue ben, ar, or may be, if they were brought before him. Nothing more variable than voices, yet men can likewife difcern them perfonally, nay you fhall haue a *Buffon*, or *Pantomimus* will exprefle as many as hee pleafeth. Nothing more variable, than the differing founds of words, yet men haue found the way to re-duce thẽ to a few fimple Letters; fo that it is not the *infufficiency or incapacity of mans mind;* but it is the *re-mote ftanding or placing thereof*, that breedeth thefe Mazes and incomprehenfions; for as the fence a far off, is full of miftaking, but is exaƐt at hand, fo is it of the vnderftanding; The remedie whereof, is not to quicken or ftrengthen the Organ, but to goe nee-rer to the obieƐt; and therefore there is no doubt, but if the Phifitians will learne, and vfe the true ap-proaches and *Auenues* of Nature, they may affume as much as the Poet fayth;

    Et quoniam variant Morbi, variabimus artes,
      Mille Mali fpecies, mille Salutis erunt.

Which that they fhould doe, the noblenefle of their Art doth deferue; well fhadowed by the Po-ets, in that they made *Aefculapius* to be the fonne of Sunne, the one being the fountaine of life, the other as the fecond ftreame; but infinitely more honored by the example of our Sauiour, who made the body of man the obieƐt of his miracles, as the foule was the obieƐt of his DoƐtrine. For wee reade not that euer he vouchfafed to doe any miracle about honor,

                      or

or money, (except that one for giuing Tribute to
Cæſar) but onely about the preſeruing, ſuſtayning,
and healing the bodie of man.

Medicine is a Science, which hath beene (as wee
haue ſayd)more profeſſed, than labored,& yet more
labored, than aduanced; the labor hauing been, in
my iudgement, rather in circle,than in progreſsion.
For, I finde much Iteration, but ſmall Addition. It
conſidereth *cauſes of Diſeaſes, with the occaſions or im-
pulſions* : The *Diſeaſes themſelues*, with the *Accidents :* and the *Cures*, with the *Preſeruations.* The
Deficiences which I thinke good to note, being a
few of many,& thoſe ſuch,as ar of a more open and
manifeſt Nature,I will enumerate,and not place.

The firſt is the diſcontinuance of the auncient
and ſerious diligence of *Hippocrates*, which vſed to
ſet downe a Narratiue of the ſpeciall caſes of his pa-
tientes and how they proceeded,& how they were
iudged by recouery or death. Therefore hauing
an example proper in the father of the art, I ſhal not
neede to alledge an example forraine, of the wiſedome of the Lawyers, who are carefull to reporte
new caſes and deciſions, for the direction of future
iudgements. This continuance of *Medicinall Hiſtory*,
I find deficient, which I vnderſtand neither to be ſo
infinite as to extend to euery *common Caſe*,nor ſo reſerued, as to admit none but *Woenders* : for many
thinges are new in the *Manner*, which are not new
in the *Kinde*,and if men will intend to obſerue, they
ſhall finde much worthy to obſerue.

*Narratio-
nes Medi-
cinales.*

In the inquirie which is made by *Anatomie,* I finde much deficience : for they enquire of the *Parts,* and their *Substances, Figures,* and *Collocations;* But they enquire not of the *Diuersities of the Parts;* the *Secrecies of the Passages;* and the *seats or neastling of the humours ;* nor much of the *Foot-steps, and impressions of Diseases;* The reason of which omission, I suppose to be, because the first enquirie may be satisfied, in the view of one or a few Anatomies: but the latter being comparatiue and casuall, must arise from the view of many. And as to the diuersitie of parts, there is no doubt but the facture or framing of the inward parts, is as full of difference, as the outward, and in that, is the *Cause Continent* of many diseases, which not being obserued, they quarrell many times with the humors which are not in fault, the fault being in the very frame and Mechanicke of the parte which cannot be remoued by medicine alteratiue but must be accomodate and palliate by dyets and medicines familiar. And for the passages and pores, it is true which was aunciently noted that the more subtile of them appeare not in anatomyes, because they are shut and latent in dead bodies, though they be open and manifest in liue: which being supposed, though the inhumanity of *Anatomia viuorū* was by *Celsus* iustly reproued: yet in regard of the great vse of this obseruation the inquiry needed not by him so sleightly to haue ben relinquished altogether, or referred to the casuall pra&ctises of surgerie, but mought haue been well diuerted vpon the dissection of beastes aliue,

liue, which notwithstanding the dissimilitude of their parts, may sufficiently satisfie this inquirie. And for the humors, they are commonly passed ouer in Anatomies, as purgaments, whereas it is most necessarie to obserue, what cauities, nestes & receptacles the humors doe finde in the parts, with the differing kinde of the humor so lodged and receiued. And as for the footesteps of diseases, & their deuastations of the inward parts, impostumations, exulcerations, discontinuations, putrefactions, consumptions, contractions, extensions, convulsions, dislocations, obstructions, repletions, together with all preternatural substances, as stones, carnosities, excrescences, wormes, and the like: they ought to haue beene exactly obserued by multitude of Anatomies, and the contribution of mens seuerall experiences; and carefully set downe both historically according to the appearances, and artificially with a reference to the diseases and symptomes which resulted from them, in case where the Anatomy is of a defunct patient: wheras now vpon opening of bodies, they are passed ouer sleightly, and in silence.

In the inquirie of diseases, they doe abandon the cures of many, some as in their nature incurable, and others, as passed the periode of cure; so that Sylla and the *Triumvirs* neuer proscribed so many men to die, as they doe by their ignorant edictes, whereof numbers do escape with lesse difficulty, then they did in the Romane proscriptions. Therfore I wil not doubt, to note as a deficience, that they inquire not

Inquisitio vlterior c Morbis in sanabilibus.

the

the perfite cures of many difeafes, or extremities of difeafes, but pronouncing them incurable, doe enact a lawe of neglect, & exempt ignorance from difcredite.

De Eutha-   Nay further, I efteeme it the office of a Phifiti-
nafia exte- on, not onely to reftore health, but to mittigate pain
riore. and dolors, and not onely when fuch mittigation may conduce to recouery, but when it may ferue to make a fayre and eafie paffage : for it is no fmall fe-licitie which *Auguftus Cæfar* was wont to wifh to himfelfe, that fame *Euthanafia,* and which was fpeci-ally noted in the death of *Antoninus Pius,* whofe death was after the fafhion and femblance of a kind-ly & pleafant fleepe. So it is written of *Epicurus,* that after his difeafe was iudged defperate, he drowned his ftomacke and fenfes with a large draught and in-gurgitation of wine, whereupon the Epigram was made; *Hinc ftygias Ebrius haufit aquas:* He was not fober enough to tafte any bitterneffe of the ftygian water. But the Phifitions contrariwife doe make a kinde of fcruple and Religion to ftay with the pati-ent after the difeafe is deplored, wheras in my iudg-ment they ought both to enquire the skill, and to giue the attendances for the facilitating & affwaging of the paynes and agonies of death.

Medicinæ   In the confideration of the Cures of difeafes, I find
experi- a deficience in the Receiptes of proprietie, refpec-
mentales. ting the particular cures of difeafes : for the Phifiti-ans haue fruftrated the fruite of tradition & experi-ence by their magiftralities, in adding and taking out

and

and changing, *Quid pro quo,* in their receiptes, at
their pleasures, commanding so ouer the medicine,
as the medicine cannot commmād ouer the disease:
For except it be Treacle and Mythridatū, & of late
Diascordium, and a few more, they tye themselues to
no receiptes seuerely and religiously: for as to the
confections of sale, which are in the shoppes, they
are for readines, and not for proprietie: for they are
vpon generall intentions of purging, opening, comforting, altering, and not much appropriate to particular Diseases; and this is the cause why Emperiques, and ould women are more happie many
times in their Cures, than learned Phisitians; because they are more religious in holding their Medicines. Therefore here is the deficience which I
finde, that Phisitians haue not partly out of their
owne practize, partly out of the constant probations
reported in bookes; & partly out of the traditions of
Emperiques: set downe and deliuered ouer, certaine
Experimentall Medicines, for the Cure of particular
Diseases; besides their owne *Coniecturall* and *Magistrall descriptions.* For as they were the men of the
best Composition in the State of *Rome,* which either
being Consuls inclined to the people; or being Tri-
bunes inclined to the Senat: so in the matter we now
handle, they be the best Phisitians, which being learned incline to the traditions of experience; or being *Imitatio*
Emperiques, incline to the methods of learning. *Naturæ in*

   In preparation of Medicines, I doe finde strange *Balneis, &*
specially, considering how mineral Medicines haue *Aquis Me-*
beene *dicinalibus*

beene extolled ; and that they are fafer, for the outward, than inward parts, that no man hath fought,
to make an Imitation by Art of Naturall Bathes, and
Medicinable fountaines: which neuerthelefle are
confeſſed to receiue their vertues from Minerals:
and not ſo onely, but difcerned and diftinguiſhed
from what particular Mynerall they receiue Tincture, as Sulphur, Vitriole, ſteele, or the like: which
Nature if it may be reduced to compoſitions of art,
both the varietie of them will be encreaſed, & the
temper of them will be more commanded.

Filum Me-   But leaſt I grow to be more particular, than is a-
dicinale, ſi- greeable, either to my intention, or to proportion, I
ue de vici- will conclude this part with the note of one defici-
bus Medi- ence more, which ſeemeth to me of greateſt confe-
cinarum. quence, which is, that the preſcripts in vſe, are too
compendious to attaine their end; for to my vnderſtanding, it is a vaine and flattering opinion, to think
any Medicine can be ſo foueraigne, or ſo happie, as
that the Receit or vſe of it, can worke any great effect vpon the bodie of man; it were a ſtrange ſpeach,
which ſpoken, or ſpoken oft, ſhould reclaime a man
from a vice, to which he were by nature ſubiect:
it is order, pourſuite, fequence, and interchange of
application, which is mightie in nature ; which although it require more exact knowledge in preſcribing, and more precife obedience in obſeruing,
yet is recompenced with the magnitude of effects.
And although a man would thinke by the dayly viſitations of the Phiſitians, that there were a

pourſuance

pourfuance in the cure; yet let a man look into their
prefcripts and miniftrations, and he fhall finde them
but inconftancies, and euerie dayes deuifes, without
any fetled prouidence or proiect; Not that euerie
fcrupulous or fuperftirious prefcript is effectuall, no
more than euerie ftraight way, is the way to heauen,
but the *truth of the direction*, muft precede *feueritie
of obferuance.*

For *Cofmetique*, it hath parts Ciuile, and parts Ef-
feminate: for cleaneffe of bodie, was euer elteemed
to proceede from a due reuerence to God, to focietie, and to our felues. As for artificiall decoration, it
is well worthy of the deficiences which it hath : being neither fine inough to deceiue, nor handfome to
vfe, nor wholefome to pleafe.

For *Athletique*, I take the fubiect of it largely; that
is to fay, for any point of abilitie, whereunto the bodie of man may be brought, whether it be of *Actiuitie*, or of *Patience*, wherof *Actiuitie* hath two parts,
Strength and *Swiftneffe* : And *Patience* likewife hath
two parts, *Hardneffe againft wants and extremities ;*
and *Indurance of payne*, or torment ; whereof we fee
the practifes in Tumblers, in Sauages, and in thofe
that fuffer punifhment: Nay, if there be any other
facultie, which falles not within any of the former
diuifions, as in thofe that diue, that obtaine a ftrange
power of contayning refpiration, and the like, I referre it to this part. Of thefe thinges the practifes
are knowne: but the Philofophie that concerneth
them is not much enquired : the rather I thinke,

becaufe they are fuppofed to be obtayned, either by an aptneffe of Nature, which cannot be taught; or onely by continuall cuftome; which is foone prefcribed; which though it bee not true : yet I forbeare to note any deficiences : for the Olympian Games are downe long fince : and the mediocritie of thefe thinges is for vfe : As for the excellencie of them, it ferueth for the moft part, but for meı cenary oftentation.

For *Arts of pleafure fenfuall,* the chiefe deficience in them, is of *L*awes to repreffe them. For as it hath beene well obferued, that the Arts which florifh in times, while vertue is in growth, are *Militarie :* and while vertue is in State are *Liberall :* and while vertue is in declination, are *voluptuarie :* fo I doubt, that this age of the world, is fomewhat vpon the defcent of the wheele; with Arts *voluptuarie,* I couple practifes *Iocularie;* for the deceiuing of the fences, is one of the pleafures of the fences. As for Games of recreation, I hould them to belong to Ciuile life. and education. And thus much of that particular Hvmane Philosophie, Which Concernes *T* he Bodie, which is but the *T* abernacle of the minde.

For Hvmane Knovvledge, Which Concernes *T* he *M* ind, it hath two parts, the one that enquireth of The Svbstance, Or Natvre Of The Sovle Or Mind; The other, that enquireth of the Facvlties Or Fvncti-

FVNCTIONS THEREOF: vnto the firſt of theſe, the conſiderations of the *Originall of the ſoule*, whether it be *Natiue or aduentiue;* and *how farre it is exemptedfrom Lawes of Matter*; and of the *Immortalitie thereof;*and many other points do appertaine,which haue been not more laboriouſly enquired, than variouſly reported; ſo as the trauaile therein taken, ſeemeth to haue ben rather in a Maze,than in a way. But although I am of opinion,that this knowledge may be more really and ſoundly enquired euen in Nature, than it hath been; yet I hold, that in the end it muſt be bounded by Religion ; or elſe it will bee ſubieƈt to deceite and deluſion : for as the ſubſtance of the ſoule in the Creation, was not extraƈted out of the Maſſe of heauen and earth, by the benediƈtion of a *Producat :* but was immediately inſpired from God ; ſo it is not poſsible that it ſhould bee (otherwiſe than by accident) ſubieƈt *to the Lawes of Heauen and Earth;* which are *the ſubieƈt of Philoſophie;* And therefore the true knowledge of the Nature, and ſtate of the ſoule, muſt come by the ſame inſpiration, that gaue the ſubſtance. Vnto this part of knowledge touching the ſoule,there be two appendices, which as they haue ben handled, haue rather vapoured foorth fables, than kindled truth; D I V IN A T I O N, and F A S C I N A T I O N.

D I V I N A T I O N, hath beene anciently and fitly diuided into *Artificiall and Naturall ;* whereof *Artificiall* is, when the minde maketh a prediƈtion by argument, concluding vpon ſignes and tokens: *Na-*
M m *turall*

turall is, when the minde hath a prefention by an internall power, without the inducement of a figne. *Artificiall* is of two forts, either when the argument is coupled with a deriuation of caufes, which is *rationall*; or when it is onely grounded vpon a Coincidence of the effect, which is *experimentall;* whereof the later for the moft part, is fuperftitious: Such as were the Heathen obferuations, vpon the infpection of Sacrifices, the flights of birds, the fwarming of Bees; and fuch as was the *Chaldean Aftrologie*, and the like. For *Artificall Diuination*, the feuerall kinds thereof are diftributed amongft particular knowledges. The *Aftronomer* hath his predictions, as of coniunctions, afpects, Eclipfes, and the like. The Philitian hath his predictions, of death, of recouerie, of the accidents and iffues of Difeafes. The Politique hath his predictions; *O vrbem venalem, & cito perituram, fi emptorem inuenerit ;* which ftayed not long to bee perfourmed in *Sylla* firft, and after in & *Cæfar.* So as thefe predictions are now impertinēt, to be referred ouer. But the *Diuination*, which fpringeth frō the internal nature of the foul, is that which we now fpeak of which hath ben made to be of two forts; *Primitiue* and by *Influxion*. Primitiue is grounded vpon the fuppofition, that the minde when it is withdrawne and collected into it felfe, and not diffufed into the Organes of the bodie, hath fome extent and latitude of prenotion; which therefore appeareth moft in fleepe, in extafies, and nere death; and more early in waking apprehenfions; and is induced

duced and furthered by thofe abſtinences, and ob-
ſeruances, which make the minde moſt to conſiſt in
it ſelfe. By influxion is grounded vpon the conceit,
that the mind, as a mirror or glaſſe, ſhould take illumination from the fore knowledge of God and ſpirits, vnto which the ſame Regiment doth likewiſe
conduce. For the retyring of the minde within it
ſelfe, as the State which is moſt ſuſceptible of diuine influxions; ſaue that it is accompanied in this
caſe with a feruencie and eleuation, (which the ancients noted by furie) and not with a repoſe and
and quiet, as it is in the other.

Faſcination is the power and aĉt of Imagination,
intenſiue vpon other bodies, than the bodie of the
Imaginant; for of that we ſpake in the proper place:
wherein the Schoole of *Paracelſus*, and the Diſciples
of pretended Naturall *Magicke*, haue beene ſo in-
temperate, as they haue exalted the power of the
imagination, to be much one with the power of
Miracle-working faith : others that drawe neerer
to Probabilitie, calling to their view the ſecret paſſages of things, and ſpecially of the Contagion that
paſſeth from bodie to bodie, doe conceiue it ſhould
likewiſe be agreeable to Nature, that there ſhould
be ſome tranſmiſsions and operations from ſpirit to
ſpirit, without the mediation of the ſences, whence
the conceits haue growne, (now almoſt made ciuile)
of the Maiſtring Spirite, & the force of confidence,
and the like. Incident vnto this, is the inquirie how
to raiſe and fortifie the imagination, for if the Imagination

gination fortified haue power, then it is materiall to know how to fortifie and exalt it. ·And herein comes in crookedly and dangeroufly, a palliation of a great part of *Ceremoniall Magicke*. For it may bee pretended, that *Ceremonies, Charaĉters*, and *Charmes* doe worke, not by any *Tacite* or *Sacramentall contraĉt* with euill fpirits; but ferue onely to ftrengthen the imagination of him that vfeth it; as Images are faid by the *Romane Church*, to fix the cogitations, and raife the deuotions of them that pray before them. But for mine owne iudgment, if it be admitted that Imagination hath power; and that *Ceremonies* fortifie Imagination, & that they be vfed fincerely & intentionally for that purpofe: yet I fhould hold them vnlawfull, as oppofing to that firft ediĉt, which God gaue vnto man. *In fudore vultus comedes Panem tuum*. For they propound thofe noble effeĉts which God hath fet foorth vnto man, to bee bought at the price of Laboure, to bee attained by a fewe eafie and flothful obferuances. Deficiences in thefe knowledges I wil report none, other than the generall deficience, that it is not knowne, how much of them is veritie, and how much vanitie.

THE KNOVVLEDGE WHICH RESPECTETH THE FACVLTIES OF THE MINDE OF MAN, is of two kinds: The one refpeĉting his VNDERSTANDING and REASON, and the other his WILL, APPETITE, & AFFECTION, wherof the former produceth POSITION or DECREE, the later ACTION or EXECVTION. It is true that the *Imagination* is an *Agent*, or *Nuntius*

tius in both Prouinces, both the *Iudiciall,* and the *Ministeriall.* For *Sence* sendeth ouer to *Imagination,* before *Reason* haue iudged: and *Reason* sendeth ouer to *Imagination,* before the *Decree* can be acted. For *Imagination* euer precedeth *Voluntary Motion.* Sauing that this *Ianus* of *Imagination* hath differing faces; for the face towards *Reason,* hath the print of Truth. But the face towards *Action,* hath the print of *Good;* which neuerthelesse are faces,

Quales decet esse sororum. Neither is the *Imagination* simply and onely a Messenger; but is inuested with, or at least wise vsurpeth no small autho- no small authoritie in it selfe; besides the duty of the Message. For it was well sayd by *Aristotle : That the minde hath ouer the Bodie that commaundement which the Lord hath ouer a Bond-man ; But, that Reason hath ouer the Imagination that Commandement, which a Magistrate hath ouer a free Citizen;* who may come also to rule in his turne. For we see, that in matters of *Faith* & *Religion,* we raise our *Imagination* aboue our *Reason,* which is the cause why *Religion* sought euer accesse to the *Minde* by *Similitudes,* *Types,* *Parables,* *Visions,* *Dreames.* And againe in all perswasions that are wrought by eloquence, and o-the impression of like Nature, which doe paint and disguise the true appearance of thinges, the cheefe recommendation vnto *Reason,* is from the *Imagination.* Neuerthelesse, because I finde not any Science, that doth properly or fitly pertaine to the IMAGINATION, I see no cause to alter the former diuision. For as for Poesie, it is rather a pleasure, or play of imagination, than a worke or dutie thereof. And if it be a worke; wee speake not nowe of such partes of learning, as the Imagination produceth, but of such Sciences, as handle and consider of the *Imagination.* No more than wee shall speake nowe of such *Knowledges,* as reason produceth, (for that extendeth to all Philosophy) but of such *Knowledges,* as doe handle and enquire of the facultie of *Reason* ; So as *Poesie* had his true place. As for the power of the *Imagination* in nature, and the manner of fortifying the same, wee haue mentioned it in the Doctrine *De Anima,* whervnto most fitly it belongeth. And lastly, for *Imaginatiue,* or *Insinuatiue Reason,* which is the subiect of Rhetoricke, wee thinke it best to referre it to the *Arts of Reason.* So therefore we content our selues with the former diuision, that Humane Philosophy, which respecteth the faculties of the minde of man, hath two parts, R A T I O N A L L and *M O R A L L.*

The part of humane *Philosophie,* which is Rationall, is of all knowledges, to the most wits, the least delightfull : and seemeth but a Net of subtilitie and spinositie. For as it was truely sayd, that Knowledge is *Pabulum animi* ; So in the Nature of mens appetite to this foode, most men are of the tast and stomach of the Israelites in the desert, that would faine haue returned *Ad ollas carnium,* and were wearie of *Manna,* which though it were celestiall, yet seemed lesse nutritiue and comfortable. So generally men tast well

well knowledges that are drenched in fleſh and blood, *Ciuile Hiſtorie, Mortlitie, Policie,* about the which mens affeℭions praiſes, fortunes doe turne and are conuerſant:But this ſame *Lumen ſiccum,*doth parch and offend moſt mens watry and ſoft natures. But to ſpeake truly of thinges as they are in worth, R A T I O N A L L *Knowledges;* are the keyes of all other Arts; For as *Ariſtotle* ſayth aptly and elegantly, *That the hand is the Inſtrument of Inſtruments; and the miude is the Fourme of Fourmes:* So theſe be truely ſaid to be the Art of Arts: Neither do they onely direℭ, but likewiſe confirme and ſtrengthen : euen as the habite of ſhooting, doth not onely inable to ſhoote a neerer ſhoote, but alſo to draw a ſtronger Bowe.

The A R T S I N T E L L E C T V A L L, are foure in number, diuided according to the ends whereunto they are referred : for mans labour is to *inuent* that which is *ſought* or *propounded* : or to *iudge* that which is *inuented :* or to *retaine* that which is *iudged :* or to *deliuer* ouer that which is *retained.* So as the Arts muſt bee foure : A R T S of E N Q V I R I E or I N V E N T I O N : A R T of E X A M I N A T I O N or I V D G E M E N T : A R T of C V S T O D I E or M E-M O R I E : and A R T of E L O C V T I O N or T R A-D I T I O N.

I N V E N T I O N is of two kindes much diffe-ring; The one of A R T S and S C I E N C E S, and the other of S P E E C H and A R G V M E N T S. The for-mer of theſe, I doe report deficient : which ſeemeth to me to be ſuch a deficience, as if in the making of an Inuentorie, touching the State of a defunct, it should be set downe, *That there is no readie money.* For as money will fetch all other commodities ; so this knowledge is that which should purchase all the rest. And like as the *West Indies* had neuer been discouered, if the vse of the Mariners Needle, had not been first discouered; though the one bee vast Regions, and the other a small Motion. So it cannot be found strange, if *S*ciences bee no further discodered, if the Art it selfe of *Inuention* and *Discouerie,* hath been passed ouer.

That this part of Knowledge is wanting, to my Iudgement, standeth plainely confessed : for first *Logicke* doth not pretend to inuent *Sciences,* or the *Axiomes* of *Sciences,* but passeth it ouer with a *Cuiq; in sua arte credendum.* And *Celsus* acknowledgeth it grauely, speaking of the Empirical and Dogmaticall *S*ects of Phisitians, *That Medicines and Cures, were first found out, and then after the Reasons and causes were discoursed : and not the Causes first found out, and by light from them the Medicines and Cures discouered.* And *Plato* in his *Theætetus* noteth well, *That particulars are infinite, and the higher generalities giue no sufficient direction : and that the pythe of all Sciences, which maketh the Arts-man differ from the inexpert, is in the middle propositions, which in euerie particular knowledge are taken from Tradition & Experience.* And therefore wee see, that they which discourse of the Inuentions and Originals of thinges, referre them rather to *Chaunce,* than to *Art,* and rather to *Beasts, Birds,*

Birds, Fishes, Serpents, than to *Men.*

> *Dictamnum genetrix Cretæa carpit ab Ida,*
> *Puberibus caulem folijs, & flore comantem*
> *Purpureo: non illa feris incognita Capris,*
> *Gramina cum tergo volucres hæsere sagittæ.*

So that it was no maruaile, (the manner of *Anti-quitie* being to confecrate Inuentors)that the *Ægyptians* had fo few humane Idols in their Temples, but almoft all Brute:

> *Omnigenumqne Deum monstra, & latrator Anubis*
> *Contra Neptunũ & Venerem, contraq; Mineruam &c.*

And if you like better the tradition of the Greci-ans, and afcribe the firft Inuentions to Men, yet you will rather beleeue that *Prometheus* firft ftroake the flints, and maruailed at the fparke, than that when he firft ftroke the flints, he xpected the fparke; and therefore we fee the *West Indian Prometheus,* had no intelligence with the *Europæan,* becaufe of the rarenesse with them of flint, that gaue the firft occa-fion: fo as it fhould feeme, that hetherto men are ra-ther beholden to a wilde Goat for Surgerie, or to a Nightingale for Mufique, or to the *Ibis* for fome part of Phificke, or to the Pot lidde, that flew open for Artillerie, or generally to *Chaunce,* or any thinge elfe, than to *Logicke* for the Inuention of Arts and Sciences. Neither is the fourme of Inuention, which *Virgill* defcribeth much other.

> *Vt varias vfus meditando extunderet artes,*
> *Paulatim,*

For if you obferue the words well, it is no other

methode

methode, than that which brute Beasts are capable of, and doe put in vre; which is *a perpetuall intending or practising some one thing vrged and imposed, by an absolute necessitie of conseruation of being*; For so *Cicero* sayth verie truly; *Vsus vni rei deditus, & Naturam & Artem sæpe vincit* : And therefore if it bee sayd of Men,

Labor omnia vincit
Improbus, & duris vrgens in rebus egestas;
It is likewise sayd of beasts, *Quis Psittaco docuit suum χαιρε?* who taught the Rauen in a drowth to throw pibbles into an hollow tree, where shee spyed water, that the water might rise, so as shee might come to it? who taught the Bee to sayle through such a vast Sea of ayre, and to finde the way from a field in flower, a great way off, to her Hiue? who taught the Ant to bite euerie graine of Corne, that she burieth in her hill, least it should take roote and growe? Adde then the word *Extundere*, which importeth the extreame difficultie, and the word *Paulatim*, which importeth the extreame slownesse; and we are where we were, euen amongst the *Egyptians Gods*; there being little left to the facultie of *Reason*, and nothing to the dutie of *Art* for matter of *Inuention*.

Secondly, the Induction which the *Logitians* speake of, and which seemeth familiar with *Plato*, whereby the *Principles* of *Sciences* may be pretended to be inuented, and so the middle propositions by deriuation from the Principles; their
fourme

fourme of Induction, I say is vtterly vitious and in-
competent: wherein their errour is the fowler, becaufe it is the duetie of *Art* to perfecte and
exalt Nature: but they contrariewife haue wronged, abufed, and traduced Nature. For hee that
fhall attentiuely obferue howe the minde doth gather this excellent dew of Knowledge, like vnto
that which the Poet fpeaketh of *Aerei mellis cæle-*
ftia dona, deftilling and contryuing it out of particulars naturall and artificiall, as the flowers of
the field and Garden: fhall finde that the mind
of her felfe by Nature doth mannage, and
Acte an Induction, much better than they
defcribe it. For to conclude *vppon an Enuume-*
ration of particulars without inftance contradictorie:
is no conclufion: but a coniecture; for who can
affure (in many fubiects) vppon thofe particulars,
which appeare of a fide, that there are any other
on the contrarie fide, which appeare not? As if
Samuell fhould haue refted vppon thofe Sonn s
of *Iffay*, which were brought before him, and
fayled of *Dauid*, which was in the field. And this
fourme (to fay truth) is fo groffe: as it had not
beene pofsible for wittes fo fubtile, as haue mannaged thefe thinges, to haue offered it to the
world, but that they hafted to their *Theories* and
Dogmaticals, and were imperious and fcornefull
toward particulars, which their manner was to
vfe, but as *Lictores* and *Viatores* for Sargeants
and Wifflers, *Ad fummonendam turbam*, to make

Nn 2   way

way and make roome for their opinions, rather than in their true vſe and ſeruice; certainely, it is a thing may touch a man with a religious woonder, to ſee how the foot ſteps of ſeducement, are the very ſame in Diuine and Humane truth : for as in Diuine truth, Man cannot endure to become as a Child; So in Humane, they reputed the attending the Induⱪions (whereof wee ſpeake) as if it were a ſecond Infancie or Child hood.

Thirdly, allowe ſome *Principles* or *Axiomes* were rightly induced; yet neuertheleſſe certaine it is, that *Middle Propoſitions*, cannot be diduced from them in *Subiect of Nature* by *Syllogiſme*, that is, *by Touch and Reduction of them to Principles in a Middle Terme*. It is true, that in Sciences popular, as *Moralities, Lawes,* and the like, yea, and *Diuinitie* (becauſe it pleaſeth God to apply himſelfe to the capacity of the ſimpleſt) that fourme may haue vſe; and in *Naturall Philoſophie* likewiſe, by way of argument or ſatisfactorie *Reaſon, Quæ aſſenſum parit, Operis Effœta eſt* : But the ſubtiltie of Nature and Operations will not bee inchayned in thoſe bonds: For *Arguments* conſiſt of *Propoſitions*, and *Propoſitions* of *Words*, and *Wordes* are but the *Current Tokens or Markes of popular Notions of things* : which Notions if they bee groſſely and variably collected out of Particulars; It is not the laborious examination either *of Conſequences of Arguments*, , *or of the truth of Propoſitions*, that can euer correct that Errour; being (as the Phi-

ſitians

fitians fpeake) in the firft digeftion ; And therefore
it was not without caufe, that fo many excellent
Philofophers became *Sceptiques* and *Academiques*,
and denyed any certaintie of Knowledge, or
Comprehenfion, and held opinion that the knowledge of man extended onely to Appearances, and
Probabilities. It is true, that in *Socrates* it was fuppofed to be but a fourme of *Irony, Scientiam diſſimu-
lando ſimulauit*: For hee vfed to difable his knowledge, to the end to inhanfe his Knowledge, like
the Humor of *Tiberius* in his beginnings, that would
Raigne, but would not acknowledge fo much;
And in the later *Academy*, which *Ciccro* embraced ; this opinion alfo of *Acatalepſia* (I doubt) was
not held fincerely ; for that all thofe which excelled
in Copie of fpeech, feeme to haue chofen that
Sect, as that which was fitteft to giue glorie
to their eloquence, and variable difcourfes : being rather like Progrefles of pleafure, than
Iourneyes to an end. But affuredly many fcattered in both *Academyes*, did hold it in fubtiltie,
and integritie. But heere was their cheefe Errour ; They charged the deceite vppon the
THE SENCES; which in my Iudgement (notwithftanding all their Cauillations) are verie fufficient to certifie and report truth (though not
alwayes immediately, yet by comparifon ; by
helpe of inftrument ; and by producing, and vr-
ging fuch things, as are too fubtile for the fence, to
fome effect comprehenfible, by the fence, and other

like afsiſtáce. But they ought to haue charged the deceit *vpon the weaknes of the intellectual powers, & vpon the maner of collecting, & concluding vpon the reports of the fences.* This I ſpeake not to diſable the minde of man, but to ſtirre it vp to ſeeke helpe: for no man, be he neuer ſo cunning or practiſed, can make a ſtraight line or perfect circle by ſteadineſſe of hand, which may bee eaſily done by helpe of a Ruler or Compaſſe.

Experientia literata, & interpretatio Naturæ. This part of *Inuention,* concerning the *Inuention* of *Sciences*, I purpoſe (if God giue mee leaue) hereafter to propound : hauing digeſted it into two partes : whereof the one I tearme *Experientia literata*, and the other *Interprætatio Naturæ* : The former, being but a degree and rudiment of the later. But I will not dwell too lgng, nor ſpeake too great vpon a promiſe.

The *Inuention* of ſpeech or argument' is not properly an *Inuention* : for to *Inuent* is to diſcouer that we know not, & not to recouer or reſūmon that which wee alreadie knowe ; and the vſe of this *Inuention*, is no other ; *But out of the Knowledge, whereof our minde is alreadie poſſeſt, to drawe foorth, or call before vs that which may bee pertinent to the purpoſe, which wee take into our conſideration.* So as to ſpeake truely, it is no *Inuention* ; but a *Remembrance* or *Suggeſtion*, with a Application : Which is the cauſe why the *Schooles* doe place it after Iudgement, as ſubſequent and not precedent. Neuertheleſſe, becauſe wee doe account it a Chaſe, aſwell afwell of Deere in an inclofed Parke, as in a Forreft at large : and that it hath alreadie obtayned the the name : Let it bee called *Inuention* ; fo as it be perceyued and difcerned, that the *S*cope and end of this *Inuention*, is readyneffe and prefent vfe of our knowledge, and not addition or amplification thereof.

To procure this readie vfe of Knowledge, there are two Courfes : PREPARATION and SVG-GESTION. The former of thefe, feemeth fcarcely a part of Knowledge ; confifting rather of Diligence, than of any artificiall erudition. And heerein *Ariftotle* wittily, but hurtfully doth de-ride the *Sophifts,* neere his time, faying, *They did as if one that profeffed the Art of Shooe-making, fhould not teach howe to make vp a Shooe, but onely ex-hibite in a readineffe a number of Shooes of all fafhions and Sizes.* But yet a man might reply, that if a *S*hooe-maker fhould haue no Shooes in his *S*hoppe, but onely worke, as hee is befpoken, hee fhould bee weakely cuftomed. But our *S*auiour, fpeaking of Diuine Knowledge, fayth : *That the Kingdome of Heauen, is like a good Houfholder, that bringeth foorth both newe and ould ftore :* And wee fee the ancient Writers of *Rhetoricke* doe giue it in precept : **That Pleaders fhould haue the Places, whereof they haue moft continuall vfe, readie handled in all the barietie that may bee,** as that, **To fpeake fo? the literall Interp?etation of the Law againft Equitie,** and **Contrarie : and to fpeake fo? P?e-fumptions**

fumptions and Inferences againſt Teſtimonie ; and
Contrarie: And *Cicero* himſelfe, being broken vnto
it by great experience, deliuereth it plainely ; That
whatſoeuer a man ſhall haue occaſion to ſpeake of,
(if hee will take the paines) he may haue it in effeᶜt
premeditate, and handled in theſe.   So that when
hee commeth to a particular, he ſhall haue nothing
to doe, but to put too Names, and times, and places;
and ſuch other Circumſtances of Indiuiduals. We
ſee likewiſe the exaᶜt diligence of *Demoſthenes*, who
in regard of the great force, that the entrance and
acceſſe into cauſes hath to make a good impreſsion ;
had readie framed a number of *Prefaces* for Orations and Speeches. All which Authorities and Pre-
ſidents may ouer-way *Ariſtotles* opinion, that would
haue vs chaunge a rich Wardrobe, for a paire of
Sheares.

    But the Nature of the Colleᶜtion of this *Prouiſi-
on* or *Preparatorie ſtore*, though it be common, both
to *Logicke*, and *Rhetoricke* ; yet hauing made an en-
trye of it heere, where it came firſt to be ſpoken of ;
I thinke fitte to referre ouer the further handling of
it to *Rhetoricke.*

    The other part of INVENTION, which I terme
SVGGESTION, doth aſsigne and direᶜt vs to
certaine *Markes* or *Places*, which may excite our
Minde to returne and produce ſuch Knowledge, as
it hath formerly colleᶜted : to the end wee may
make vſe thereof. Neither is this vſe (truely taken)
onely to furniſh argument, to diſpute probably
                                                             with

with others ; But likewife to Minifter vnto our
Iudgement to conclude aright within our felues.
Neither may thefe places ferue onely to apprompt
our Inuention; but alfo to direct our enquirie. For
a facultie of wife interrogating is halfe a knowledge;
For as *Plato* faith;*Whofoeuer feeketh,knoweth that which
he feeketh for,in a generall Notion ; Elfe how fhall he know
it, when he hath found it?*And therfore the larger your
Anticipation is, the more direct and compendious is
your fearch. But the fame *Places* which will help vs
what to produce, of that which we know alreadie ;
will alfo helpe vs, if a man of experience were before vs,what queftions to aske;or if we haue Bookes
and Authors, to inftruct vs what points to fearch and
reuolue: fo as I cannot report,that this part of *Inuention*, which is that which the Schooles call *Topiques*,
is deficient.

   Neuertheles *Topiques* are of 2. forts *general* & *fpeciall.* The *generall* we haue fpoke to; but the particular
hath ben touched byfome,but reiected generally,as
inartificial & variable. But leauing the humor which
hath raigned too much in the Schooles (which is to
be vainly fubtile in a few thinges, which are within
their command,and to reiect the reft) I doe receiue
particular *Topiques*, that is places or directions of
Inuention and *Inquirie* in euery particular knowledg,
as thinges of great vfe; being Mixtures of *Logique*
with the Matter of *Sciences*: for in thefe it holdeth;
Ars inueniendi adolefcit cum Inuentis : for as in going of a way, wee doe not onely gaine that part

of the waye which is paffed, but wee gaine the better light of that part of the waye which remayneth: So euerie degree of proceeding in a Science giueth a light to that which followeth ; which light if wee ftrengthen, by drawing it foorth into queftions or places inquirie, wee doe greatly aduance our pourfuyte.

Nowe wee paffe vnto the ARTES OF IVDGE-MENT, which handle the Natures of *Proofes* and *Demonftrations*; which as to *Induction* hath a Coincidence with *Inuention* : For all *Inductions whether in good or vitious fourme, the fame action of the Minde which Inuenteth, Iudgeth ; all one as in the fence* : But otherwife it is in proofe by *Syllogifme* : For the proofe beeing not immediate but by meanes the *Inuention of the Meane* is one thinge : and the *Iudgement of the Confequence* is another. The one *Excyting* onely: the other *Examining* : Therefore for the reall and exacte fourme of Iudgement, wee referre our felues to that which wee haue fpoken of *Interpretation of Nature.*

For the other Iudgement by *Syllogifme*, as it is a thinge moft agreeable to the Minde of Man : So it hath beene vehementlye and excellently laboured. For the Nature of Man doth extreamelye couet, to haue fomewhat in his Vnderftanding fixed and vnmooueable, and as a Reft, and Support of the Mind. And therefore as *Ariftotle* endeuoureth to prooue, that in all Motion, there is
                                                              fome

some pointe quiescent; and as hee elegantlye expoundeth the auncient Fable of *Atlas*, (that stood fixed, and bare vp the Heauen from falling) to bee meant of the Poles or Axel tree of Heauen, wherevppon the conuersion is accomplished ; so assuredlye men haue a desire, to haue an *Atlas* or Axel tree within : to keepe them from fluctuation ; which is like to a perpetuall perill of falling: Therefore men did hasten to sette downe some Principles, about which the varietie of their disputations might turne.

So then this Art of I V D G E M E N T, is but the *Reduction* of *Propositions*, to *Principles* in a *Middle Tearme.* The *Principles* to bee agreed by all, and exempted from Argument ; The M I D D L E T E A R M E to bee elected at the libertie of euerie Mans *Inuention*: The *Reduction* to be of two kindes *Direct*, and *Inuerted*; the one when the *Proposition* is reduced to the *Principle*, which they terme a *Probation ostensiue* : the other when the contradictorie of the Proposition is reduced to the contradictorie of the Principles, which is, that which they call *Per Incommodum*, or *pressing an absurditie*: the *Number* of *Middle Termes* to be, as the *Proposition* standeth, *Degrees* more or lesse, remooued from the *Principle*.

But this Arte hath twoo seuerall Methodes of Doctrine : the one by way of *Direction*, the other by way of *Caution* : the former frameth and setteth downe *a true Fourme of Consequence*, by the

variations

variations and deflexions, from which Errours
and Inconfequences may bee exactly iudged. Toward the Compofition and ftructure of which
fourme, it is incident to handle the partes thereof,
which are *Propofitions*, and the partes of *Propofitions*, which are SIMPLE WORDES. And this
is that part of *Logicke*, which is comprehended in the
Analytiques.

The fecond Methode of Doctrine, was introduced for expedite vfe, and affurance fake ; difcouering the more fubtile fourmes of *Sophifmes*, and
Illaqueations, with their *redargutions*, which is that
which is tearmed ELENCHES. For although in
the more groffe fortes of Fallacies it happeneth (as
Seneca make the comparifon well) as in iugling
feates, which though wee knowe not howe they
are done; yet wee knowe well it is not, as it feemeth to bee : yet the more fubtile fort of them
doth not onely put a man befides his anfwere, but
doth many times abufe his Iudgment.

This part concerning ELENCHES, is excellently handled by *Ariftotle* in *Precept*, but more
excellently by *Plato in Example* : not onely in the
perfons of the *Sophifts*, but euen in *Socrates* himfelfe, who profefsing to affirme nothing, but to
infirme that which was affirmed by another, hath
exactly expreffed all the fourmes of obiection, fal-
lace and redargution. And although wee haue
fayd that the vfe of this Doctrine is for *Redargution*: yet it is manifeft, the degenerate and corrupt

rupt is vſe for *Caption* and *Contradiction,* which paſſeth for a great facultie, and no doubt, is of verie great aduauntage ; though the difference bee
good which was made betweene Orators and
Sophiſters, that the one is as the Greyhound,
which hath his aduauntage in the race, and the
other as the Hare, which hath her aduantage in the
turne, ſo as it is the aduauntage of the weaker creature.

But yet further, this Doctrine of E L E N-
C H E S, hath a more ample latitude and extent,
than is perceiued : namely vnto diuers partes of
Knowledge : whereof ſome are laboured, and
other omitted. For firſt, I conceiue (though it
maye ſeeme at firſt ſomewhat ſtrange) that that
part which is variably referred, ſometimes to *Logicke*, ſometimes to *Metaphyſicke*, touching the
Common aduuncts of Eſſences, is but an *Elenche*: for the
great *Sophiſme of all Sophiſmes,* beeing *Æquiuo-
cation* or *Ambiguitie of Wordes and Phraſe,* ſpecially of ſuch wordes as are moſt generall and
interueyne euerie Enquirie : It ſeemeth to mee
that the true and fruitfull vſe, (leauing vaine ſub-
tilities and ſpeculations) of the Enquirie, *Ma-
ioritie, Minoritie, Prioritie, Poſterioritie, Iden-
titie, Diuerſitie, Poſſibilitie, Acte, Totalitie,
Partes, Exiſtence, Priuation,* and the like, are
but wiſe Cautions againſte Ambiguityes of
Speech. So againe, the diſtribution of thinges
into certaine Tribes, which we call *Categories* or

Predicaments, are but Cautions againſt the confu-ſion of *Definitions* and *Diuiſions.*

Secondly, there is a ſeducement that worketh by the ſtrength of the Impreſsion, and not by the ſubtiltie of the Illaqueation, not ſo much perplex-ing the Reaſon, as ouer-ruling it by power of the *Imagination.* But this part I thinke mcre proper to handle, when I ſhall ſpeake of R H E T O-R I C KE.

But laſtly, there is yet a much more important and profound kinde of Fallacies in the Minde of Man, which I finde not obſerued or enquired at all, and thinke good to place heere, as that which of all others appertayneth moſt to rectifie I V D G E-M E N T. The force whereof is ſuch, as it doth not dazle, or ſnare the vnderſtanding in ſome par-ticulars, but doth more generally, and inwardly infect and corrupt the ſtate thereof. For the mind of Man is farre from the Nature of a cleare and equall glaſſe, wherein the beames of things ſhould reflect according to their true incidence ; Nay, it is rather like an inchanted glaſſe, full of ſuperſtition and Impoſture, if it bee not deliuered and re-duced. For this purpoſe, lette vs conſider the falſe appearances, that are impoſed vppon vs by the generall Nature of the minde, behoulding them in an example or twoo , as firſte in that inſtance which is the roote of all ſuperſtirion: Namely, *That to the Nature of the Minde of all Men it is conſonant for the Affimatiue, or Actiue to*

affect

affect, more than the negatiue or Priuatiue. So that
a fewe times hitting, or prefence, counteruayles
oft times fayling, or abfence, as was well anfwered
by *Diagoras*, to him that fhewed him in *Neptunes*
Temple, the great number of pictures, of fuch as
had fcaped Shippe-wracke, and had paide their
Vowes to *Neptune*, faying : *Aduife nowe, you that
thinke it folly to inuocate Neptune in tempeft : Yea,
but* (fayth *Diagoras*) *where are they painted that are
drowned?* Lette vs behould it in another inftance,
namely, *That the fpirite of man, beeing of an equall and
vnifourme fubftance, doth vfually fuppofe and faine in
Nature a greater equalitie and vniformitie, than is
in truth ;* Hence it commeth, that the *Mathemati-
tians* cannot fatisfie themfelues, except they reduce
the Motions of the *Celeftiall* bodyes, to perfect
Circles, reiecting fpirall lynes, and laboring to be dif-
charged of Eecentriques. Hence it commeth,
that whereas there are many thinges in Nature, as
it were *Monodica fui Iuris ;* Yet the cogitations of
Man, doe fayne vnto them *Relatiues, Parallelles,* and
Coniugates, whereas no fuch thinge is ; as they
haue fayned an Element of Fire to keepe fquare
with Earth Water, and Ayre, and the like; Nay,
it is not credible, till it bee opened, what a number of fictions and fantafies, the fimilitude of humane Actions, & Arts, together with the making of
Man *Communis Menfura,* haue brought into naturall Philofophie : not much better, than the Here-
fie of the *Anthropomorphites* bredde in the Celles

of

of groffe and folitarie *Monkes*, and the opinion of *Epicurus*, anfwearable to the fame in hea-
thenifme, who fuppofed the Gods to bee of humane Shape. And therefore *Velleius* the Epicu-
rian needed not to haue asked, why God fhould
haue adorned the Heauens with Starres, as if he
had beene an *Aedilis* : One that fhould haue
fet foorth fome magnificent fhewes or playes? for
if that great Worke mafter had beene of an Humane difpofition, hee woulde haue cafte the
ftarres into fome pleafant and beautifull workes,
and orders, like the frettes in the Roofes of
Houfes, whereas one can fcarce finde a Pofture
in fquare, or trianangle, or ftreight line amongeft
fuch an infinite numbers, fo differing an Harmonie, there is betweene the fpirite of *Man*, and
the fpirite of Nature.

Lette vs confider againe, the falfe appearances
impofed vpon vs by euerie *Mans* owne indiuiduall
Nature and Cuftome in that fayned fuppofition,
that *Plato* maketh of the Caue : for certainely, if
a childe were continued in a Grotte or *Caue*, vnder the Earth, vntill maturitie of age, and came
fuddainely abroade, hee would haue ftrange and
abfurd Imaginations; So in like manner, although
our perfons liue in the view of Heauen, yet our
fpirites are included in the *Caues* of our owne
complexions and *Cuftomes*: which minifter vnto
vs infinite Errours and vaine opinions, if they bee
not recalled to examination. But heereof wee
                                                                haue

haue giuen many examples in one of the Errors,
or peccant humours, which wee ranne briefely
ouer in our firſt Booke.

And laſtly, lette vs conſider the falſe appearances, that are impoſed vpon vs by words, which
are framed, and applyed according to the conceit,
and capacities of the Vulgar ſorte : And although
wee thinke we gouerne our wordes, and pre-
ſcribe it well. *Loquendum vt Vulgus, ſentiendum vt
ſapientes* : Yet certaine it is, that wordes, as a *Tar-
tars* Bowe, doe ſhoote backe vppon the vnderſtanding of the wiſeſt, and mightily entangle, and
peruert the Iudgement. So as it is almoſt neceſ-
ſarie in all controuerſies and diſputations, to imitate the wiſedome of the *Mathematicians*, in ſet-
ting downe in the verie beginning, the definitions
of our wordes and termes, that others may knowe
howe wee accept and vnderſtand them, and whether they concurre with vs or no. For it commeth
to paſſe for want of this, that we are ſure to end
there where wee ought to haue begun, which is in
queſtions & differences about words. To conclude
therefore, it muſt be confeſſed, that it is not poſſible
to diuorce our ſelues from theſe fallacies and falſe
appearances, becauſe they are inſeparable from our
Nature and Condition of life ; So yet neuertheleſſe the Caution of them (for all *Elenches* as
was ſaide, are but Cautions) doth extreamely
importe the true conducte of Humane Iudgement. The particular *Elenches* or *Cautions* againſt

*Elenchi
magni, ſiue
de Idolis a-
nimi hu-
mani, nati-
uis & ad-
uentitijs.*

P p   theſe

Of the aduancement of learning,

thefe three falfe appearances, I finde altogether deficient.

There remayneth one parte of Iudgement of great excellencie, which to mine vnderftanding is fo fleightly touched, as I maye reporte that alfo deficient, which is the application of the differinge kindes of Proofes, to the differing kindes of Subiects : for there beeing but foure kindes of demonftrations, that is by the immediate *confent* of the *Minde* or *Sence* ; by *Induction*; by *Sophifme*; and by *Congruitie*, which is that which *Ariftotle* calleth *Demonftration in Orbe, or Circle*, and not a *Notioribus*, euerie of thefe hath certaine Subiects in the Matter of Sciences, in which refpectiuely they haue chiefeft vfe; and certaine other, from which refpectiuely they ought to be excluded, and the rigour, and curiofitie, in requiring the more feuere Proofes in fome thinges, and chiefely the facilitie in contenting our felues with the more remiffe Proofes in others, hath beene amongeft the greateft caufes of detryment and hinderance to Knowledge. The diftributions and afsignations of demonftrations, according to the Analogie of Sciences, I note as deficient.

De Analogia Demonftrationum.

The Cuftodie or retayning of Knowledge, is either in WRITING or MEMORIE; whereof WRITINGE hath twoo partes; The Nature of the CHARACTER, and the order of the ENTRIE,

ENTRIE: for the Art of *Characters*, or other vifible notes of Wordes or thinges, it hath neereft coniugation with Grammar, and therefore I referre it to the due place ; for the *Difpofition* and *Collocation* of that Knowledge which wee preferue in Writing; It confifteth in a good Digeft of Common Places, wherein I am not ignorant of the preiudice imputed to the vfe of *Common-Place Bookes,* as caufing a retardation of Reading, and fome floth or relaxation of Memorie. But becaufe it is but a counterfeit thing in Knowledges to be forward and pregnant, except a man bee deepe and full ; I hould the Entrie of Common places, to bee a matter of great vfe and effence in ftudying ; as that which affureth copie of Inuention, and contra&eth Iudgment to a ftrength. But this is true, that of the *Methodes* of *Common places,* that I haue feen, there is none of any fufficient woorth, all of them carying meerely the face of a *Schoole*, and not of a *World*, and referring to vulgar matters, and Pedanticall Diuifions without all life, or refpe& to A&ion.

For the other Principall Parte of the Cuftodie of Knowledge, which is MEMORIE; I finde that facultie in my Iudgement weakely enquired of; An Art there is extant of it ; But it feemeth to me that there are better Precepts, than that Art ; and better pra&ifes of that Art, than thofe receiued. It is certaine, the Art (as it is) may bee rayfed to points of oftentation prodigious: But in vfe (as it is nowe mannaged) it is barrein, not burdenſome, nor dangerous to Naturall *Memorie*, as is imagined, but barren, that is, not dexterous to be ayplyed to the ſerious vſe of buſineſſe and occaſions. And therefore I make no more eſtimation of repeating a great number of Names or Wordes vppon once hearing ; or the powring ſoorth of a number of Verſes or Rimes *ex tempore* ; or the making of a *Satyricall Simile* of euerie thing, or the turning of euerie thing to a Ieſt, or the falſifying or contradicting of euerie thing by *Cauill*, or the like (wherof in the faculties of the *Minde*, there is great *Copie*, and ſuch, as by deuiſe and practiſe may bee exalted to an extreame degree of woonder ;) than I doe of the trickes of *Tumblers*, *Funambuloes*, *Baladynes* ; the one being the ſame in the *Minde*, that the other is in the bodie ; Matters of ſtrangeneſſe without worthyneſſe.

This Art of *Memorie*, is but built vpon two Intentions : The one *Prænotion* ; the other *Embleme* : *Prænotion*, diſchargeth the Indefinite ſeeking of that we would remember, and directeth vs to ſeeke in a narrowe Compaſſe : that is, ſomewhat that hath Congruitie with our *Place of Memorie*: *Embleme* reduceth conceits intellectuall to Images ſenſible, which ſtrike the *Memorie* more ; out of which *Axiomes* may bee drawne much better Practique, than that in vſe, and beſides which *Axiomes*, there are diuers moe, touching helpe of *Memorie*, not inferior to them. But I did in the beginning diſtinguiſh,

not

not to report thofe thinges deficient, which are but onely ill Managed.

There remayneth the fourth kinde of RATIO-NALL KNOVVLEDGE, which is tranfitiue, concerning the *expreßing* or *transferring* our Knowledg to others, which I will tearme by the generall name of TRADITION OR DELIVERIE. TRA-DITION hath three parres : the firft concerning the ORGANE OF TRADITION: the fecond, concerning the METHODE OF TRA-DITION : And the thirde, concerning the ILLVSTRATION OF TRADITION.

For the ORGANE OF TRADITION, it is either SPEECH OR WRITING : for *Ariftotle* fayth well : *Wordes are the Images of Cogitations, and Letters are the Images of Wordes*: But yet is not of necefsitie, that *Cogitations* bee expreffed by the *Medium of Wordes*. For *whatfoeuer is capable of fufficient differences, and thofe perceptible by the fenfe; is in Nature competent to expreße Cogitations*: And therefore we fee in the Commerce of barbarous People, that vnderftand not one anothers language, & in the practife of diuers that ar dumb & deafe, that mens minds are expreffed in geftures, though not exactly, yet to ferue the turne. And we vnderftand further, that it is the vfe of *Chyna*, and the Kingdomes of the High *Leuant*, to write in *Characters reall*, which expreffe neither *Letters, nor words in groffe*, but *Things* or *Notions*: in fo much as Countreys and Prouinces, which vnderftand not one anothers language, can neuertheleffe read one anothers Writings, becaufe

Pp 3   the

the *Characters* are accepted more generally, than the *Languages* doe extend; and and therefore they haue a vaſt multitude of *Characters*, as many (I ſuppoſe, as Radicall words.

Theſe *Notes of Cogitations* are of twoo ſortes; The one when the Note hath ſome *Similitude*, or *Congruitie* with the *Notion*; The other *Ad Placitum*, hauing force onely by *Contract* or *Acceptation*. Of the former ſort are *Hierogliphickes*, and *Geſtures*. For as to *Hierogliphickes*, (things of Ancient vſe, and embraced chiefely by the *Ægyptians*, one of the moſt ancient Nations) they are but as continued *Impreaſes* and *Emblemes*. And as for *Geſtures*, they are as *Tranſitorie Hierogliphickes*, and are to *Hierogliphickes*, as *Words ſpoken* are *to Wordes written*, in that they abide not; but they haue euermore as well, as the other an affinitie with the thinges ſignified : as *Periander* beeing conſulted with how to preſerue a tyrannie newly vſurped, bid the *Meſſenger* attend, and report what hee ſawe him doe, and went into his Garden, and topped all the higeſt flowers: ſignifying that it conſiſted in the cutting off, and keeping low of the Nobilitie and *Grandes*; *Ad Placitum*, are the *Characters reall* before mentioned, and *Words*: although ſome haue ben willing by Curious Enquirie, or rather by apt fayning, to haue deriued impoſition of Names, from Reaſon and Intendment: a ſpeculation elegant, and by reaſon it ſearcheth into *Antiquitie* reuerent : but ſparingly

mixt

mixt with truth, and of fmall fruite . This portion of knowledge, touching the *Notes of thinges*, *De Notis*
and Cogitations in generall, I finde not enquired, *Rerum.*
but deficient. And although it may feeme of no
great vfe, confidering that *Words*, and *Writings by let-*
ters, doe far excell all the other wayes : yet becaufe
this part concerneth, as it were the Mint of knowledge (for wordes, are the tokens currant and accepted for conceits, as Moneys are for values and
that it is fit men be not ignorant, that *M*oneys may
bee of another kind, than gold and filuer) I thought
good to propound it to better Enquirie.

Concerning S P E E C H and W O R D E S, the
Confideration of them hath produced the *S*ci-
ence of G R A M M A R : for *M*an ftill ftriueth to reintegrate himfelfe in thofe benedictions, from
which by his fault hee hath been depriued ; And
as hee hath ftriuen againft the firft generall *C*urfe,
by the Inuention of all other Artes : So hath hee
fought to come foorth of the feconde generall
*C*urfe, (which was the confufion of Tongues) by
the Art of G R A M M A R; whereof the vfe in another tongue is fmall : in a forreine tongue more :
but moft in fuch Forraine Tongues, as haue ceafed
to be *Vulgar Tongues*, and are turned onely to *lear-*
ned tongues. The duetie of it is of twoo Natures :
The one *Popular*, which is for the fpeedie, and perfect attayning Languages, as well for intercourfe
of Speech, as for vnderftanding of Authors : The
other *Philofophicall*, examining the power and Na-

P p 4 ture

ture of Wordes, as they are the foot ſteppes and
prints of Reaſon: which kinde of *Analogie* betweene *Wordes,* and *Reaſon* is handled *Sparſim,* bro-
kenly, though not entirely: and therefore I cannot
report it deficient, though I thinke it verie worthy
to be reduced into a Science by it ſelfe.

Vnto GRAMMAR alſo belongeth, as an Appendix, the conſideration of the Accidents of
Wordes, which are Meaſure, ſound, and Eleuation, or Accent, and the ſweeteneſſe and harſh-
neſſe of them : whence hath yſſued ſome curious obſeruations in *Rhetoricke,* but chiefely *Poeſie,*
as wee conſider it, in reſpect of the verſe, and not
of the Argument: wherein though men in learned
Tongues, doe tye themſelues to the Ancient Meaſures, yet in moderne Languages, it ſeemeth to me,
as free to make newe Meaſures of Verſes, as of
Daunces : For a Daunce is a meaſured pace, as
a Verſe is a meaſured Speech. In theſe thinges
the Sence is better Iudge, than the Art.

> *Cænæ fercula noſtræ;*
> *Mallem conuiuis, quam placuiſſe Cocis.*

And of the ſeruile expreſsing *Antiquitie* in an
vnlike and an vnfit Subiect, it is well ſayd, *Quod
tempore antiquum videtur, id incongruitate eſt maxime nouum.*

For CYPHARS; they are commonly in Letters
or Alphabets, but may bee in Wordes. *T*he
kindes

kindes of C Y P H A R S, (befides the S I M P L E
C Y P H A R S with Changes, and intermixtures
of N V L L E S, and N O N S I G N I F I-
C A N T S) are many, according to the Nature or Rule of the infoulding : W H E E L E-
C Y P H A R S, K A Y C Y P H A R S, D O V-
B L E S, &c. But the vertues of them, whereby
they are to be preferred, are three ; that they be
not laborious to write and reade ; that they bee
impofsible to difcypher ; and in fome cafes, that
they bee without fufpition. The higheft Degree whereof, is to write O M N I A   P E R
O M N I A ; which is vndoubtedly pofsible,
with a proportion Quintuple at moft, of the writing infoulding, to the writing infoulded, and
no other reftrainte whatfoeuer. This Arte of
Cypheringe, hath for Relatiue, an Art of *Difcyphe-
ringe* ; by fuppofition vnprofitable ; but, as things
are, of great vfe. For fuppofe that *Cyphars* were
well mannaged, there bee Multitudes of them
which exclude the *Difcypherer*. But in regarde
of the rawneffe and vnskilfulneffe of the handes,
through which they paffe, the greateft Matters, are many times carryed in the weakeft *Cyphars.*

In the Enumeration of thefe priuate and retyred Artes, it may bee thought I feeke to make a
greate Mufter-Rowle of Sciences ; naminge
them for fhewe and oftentation, and to little o-
ther purpofe. But lette thofe which are skilfull

in them iudge, whether I bring them in onely for apparance, or whether in that which I ſpeake of them (though in fewe Words) there be not ſome ſeede of proficience. And this muſt bee remembred, that as there bee many of great account in their Countreys and Prouinces, which when they come vp to the Seate of the Eſtate, are but of meane Ranke and ſcarcely regarded: So theſe Arts being heere placed with the principall, and ſupreame Sciences, ſeeme petty thinges: yet to ſuch as haue choſen them to ſpende their labors and ſtudies in them, they ſeeme great Matters.

For the METHODE OF TRADITION, I ſee it hath mooued a Controuerſie in our time. But as in Ciuile buſineſſe, if there bee a meeting and men fall at Wordes, there is commonly an end of the Matter for that time, and no proceeding at all: So in Learning, where there is much controuerſie, there is many times little Enquirie. For this part of knowledge of *Methode* ſeemeth to mee ſo weakely enquired, as I ſhall report it deficient.

METHODE hath beene placed, and that not amiſſe in *Logicke*, as a part of *Iudgement*; For as the Doctrine of *Syllogiſmes* comprehendeth the rules of Iudgement vppon that which is *inuented*; So the Doctrine of *Methode* contayneth the rules of *Iudgement* vppon that which is to bee deliuered,

red, for *Iudgement* precedeth *Deliuerie*, as it followeth *Inuentions*. Neither is the METHODE, or the NATVRE OF THE TRADITION materiall onely to the *Vse* of Knowledge, but likewise to the *Progress on* of Knowledge : for since the labour and life of one man, cannot attaine to perfection of Knowledge ; the *Wisedome* of the *Tradition*, is that which inspireth the felicitie of continuance, and proceding. And therefore the most reall diuersitie of *Methode*, is of METHODE REFERRED TO VSE, and METHODE REFERRED TO PROGRESSION, whereof the one may bee tearmed MAGISTRALL, and the other of PROBATION.

The later whereof seemeth to be *Via deserta & interclusa*. For as Knowledges are now deliuered, there is a kinde of Contract of Errour, betweene the Deliuerer, and the Receiuer: for he that deliuereth knowledge ; desireth to deliuer it in such fourme, as may be best beleeued ; and not as may best examined: and hee that receiueth knowledge, desireth rather present satisfaction, than expectant Enquirie, & so rather not to doubt, than not to erre: glorie making the Author not to lay open his weaknesse, and sloth making the Disciple not to knowe his strength.

But knowledge, that is deliuered as a threade to bee spunne on, ought to bee deliuered and intimated, if it were possible, *In the same Methode wherein it was inuented;* and so is it possible of know ledge induced. But in this fame anticipated and preuented knowledge; no man knoweth howe hee came to the knowledge which hee hath obtayned. But yet neuerthelefle *Secundum maius & minus,* a man may reuifite, and defcend vnto the foundations of his Knowledge and Confent : and fo tranfplant it into another, as it grewe in his owne Minde. For it is in Knowledges, as it is in Plantes ; if you meane to vfe the Plant, it is no matter for the Rootes : But if you meane to remooue it to growe, then it is more aflured to reft vppon rootes, than Slippes : So the deliuerie of Knowledges (as it is nowe vfed) is as of faire bodies of Trees without the Rootes : good for the Carpenter, but not for the Planter : But if you will haue Sciences growe ; it is lefle matter for the fhafte, or bodie of the Tree, fo you looke well to the takinge vp of the Rootes. Of which kinde of deliuerie the *Methode* of the *Mathematiques,* in that Subiect, hath fome fhadowe; but generally I fee it neither put in vre, nor put in Inquifition : and therefore note it for deficient.

De Methodo fyncera, fiue ad filios Scientiarum.

Another diuerfitie of METHODE there is, which hath fome affinitie with the former, vfed in fome cafes, by the difcretion of the Auncients; but difgraced fince by the Impoftures of many vaine perfons, who haue made it as a falfe light for their counterfeite Marchandizes ; and that is Enigmaticall and Difclofed. The pretence where-

of,

of, is to remooue the vulgar Capacities from beeing admitted to the secretes of Knowledges, and to reserue them to selected Auditors : or wittes of such sharpenesse as can pearce the vayle.

Another diuersitie of M E T H O D E, whereof the consequence is great, is the deliuerie of knowledge in A P H O R I S M E S, or in *M* E-T H O D E S ; wherein wee may obserue, that it hath beene too much taken into Custome, out of a fewe *Axiomes* or Obseruations, vppon any Subiecte, to make a solemne, and formall Art ; filling it with some Discourses, and illustratinge it with examples ; and digesting it into a sensible *Methode* : But the writinge in A P H O-R I S M E S, hath manye excellent vertues, whereto the writinge in *Methode* doth not approach.

For first, it tryeth the Writer, whether hee be superficiall or solide : For *Aphorismes*, except they should bee ridiculous, cannot bee made but of the pyth and heart of Sciences : for discourse of illustration is cut off, Recitalles of Examples are cut off : Discourse of Connexion, and order is cut off ; Descriptions of Practize, are cutte off ; So there remayneth nothinge to fill the *Aphorismes*, but some good quantitie of Obseruation : And therefore no man can suffice, nor in reason will attempt to write *Aphorismes*, but hee that is sound and grounded. But in *Methodes*,

Q q 3 *Tantum*

Of the aduancement of learning,

Tantum Series iunêturaque Pollet,
Tantum de Medio sumptis, accedit honoris :

As a Man shall make a great shew of an Art, which if it were disioynted, would come to little. Secondly, *Methodes* are more fit to winne Consent; or beleefe; but lesse fit to point to Action ; for they carrie a kinde of Demonstration in Orbe or Circle, one part illuminating another ; and therefore satisfie. But particulars beeing dispersed, doe best agree with dispersed directions. And lastlye *Aphorismes*, representing a knowledge broken, doe inuite men to enquire further ; whereas *Methodes* carrying the shewe of a Totall, doe secure men; as if they were at furthest.

Another Diuersitie of METHODE, which is likewise of great weight, is, The handling of knowledge by *Assertions*, and *their Proofes* ; or by *Questions*, and their *Determinations* : The latter kinde whereof, if it bee immoderately followed, is as preiudiciall to the proceeding of Learning, as it is to the proceedinge of an Armie, to goe about to besiege euerie little Forte, or Holde. For if the Field bee kept, and the summe of the Enterprize pursued, those smaller thinges will come in of themselues ; Indeede a Man would not leaue some important peece Enemie at his backe. In like manner, the vse of *Confutation* in the deliuerie of Sciences ought to be verie sparing;

ring; and to ferue to remooue ftronge Preoccu-
pations and Preiudgements, and not to minifter
and excite Difputations and doubts.

Another Diuerfitie of *Methodes*, is, *According
to the Subiect or Matter, which is handled.* For
there is a great difference in Deliuerie of the
Mathematiques, which are the moft abftracted of
knowledges, and *Policie*, which is the moft immerfed; And howfoeuer contention hath been mooued, touching an *vniformitie* of *Methode* in *Mul-
tiformitie* of Matter : Yet wee fee howe that opinion, befides the weakeneffe of it, hath beene
of ill defert, towardes Learning, as that which
taketh the way, to reduce Learning to certaine
emptie and barren Generalities; beeing but the verie Huskes, and Shales of Sciences, all the ker-
nell beeing forced out, and expulfed, with the
torture and preffe of the *Methode* : And therefore
as I did allow well of *particular topiques* for *Inuention* : fo I doe allow likewife of *particular Methodes
of* T*radition.*

Another Diuerfitie of *Iudgement* in the deliuerie
and teaching of knowledge, is, *According vnto the
light and prefuppofitions of that which is deliuered:* For
that knowledge, which is newe and forreine from
opinions receiued, is to bee deliuered in another
fourme, than that that is agreeable and familiar;
And therefore *Ariftotle*, when he thinkes to taxe
Democritus, doth in truth, commend him; where
hee fayth : *If wee fhall indeede difpute, and not fol-*
lowe

lowe after Similitudes, &c. For thofe, whofe conceites are feated in popular opinions, neede onely but to prooue or difpute: but thofe, whofe conceits are beyonde popular opinions, haue a double labour; the one to make themfelues conceiued, and the other to prooue and demonftrate. So that it is of necefsitie with them to haue recourfe to fimilitudes, and tranflations, to expreffe themfelues. And therefore in the Infancie of Learning, and in rude times, when thofe conceits, which are now triuiall, were then newe; the World was full of *Parables* and *Similitudes*; for elfe would men either haue paffed ouer without *Marke*, or elfe reiected for Paradoxes, that which was offered; before they had vnderftoode or iudged. So; in Diuine Learning, wee fee howe frequent *Parables* and *Tropes* are; For it is a Rule, *That whatfoeuer Science is not confonant to prefuppofitions, muft pray in ayde of Similitudes.*

There be alfo other Diuerfities of M E T H O D E S vulgar and receiued: as that of *Refolution*, or *Analyfis*, of *Conftitution*, or *Syftafis*, of *Concealement*, or *Cryptique, &c.* which I doe allowe well of; though I haue ftood vpon thofe which are leaft handled and obferued. All which I haue remembred to this

De prudentia Traditionis.

purpofe, becaufe I would erecte and conftitute one generall Enquirie (which feemes to mee deficient) touching the *Wifedome of Tradition.*

But vnto this part of Knowledge, concerning M E T H O D E, doth further belong, not onely the
Architecture

Architecture of the whole frame of a Worke, but alſo the ſeuerall beames and Columnes thereof; not as to their ſtuffe , but as to their quantitie, and figure : And therefore, *Methode* conſidereth, not onely the diſpoſition of the *Argument* or *Subiect*, but likewiſe the *Propoſitions*: not as to their *Truth* or *Matter*, but as to their *Limitation* and *Manner*. For herein *Ramus* merited better a great deale, in reuiuing the good Rules of *Propoſitions*, Καθόλυ πρωτον Κατα παντ⊙. *&c.* than he did in introducing the Canker of *Epitomes*: And yet, (as it is the Condition of Humane thinges, that according to the ancient Fables , *The moſt pretious thinges haue the moſt pernitious Keepers*) It was ſo, that the attempt of the one , made him fall vpon the other. For hee had neede be well conducted, that ſhould deſigne to make *Axiomes Conuertible*: If he make them not withall *Circular*, and *Non promouent*, or *Incurring into themſelues:* but yet the Intention was excellent.

The other Conſiderations of *Methode*, concerning *Propoſitions*, are chiefely touching the vtmoſt Propoſitions, which limit the Dimenſions of Sciences : for euerie Knowledge may bee fitly ſayd, beſides the *Profunditie* (which is the truth and ſubſtance of it , that makes it *ſolide*) to haue a *Longitude*, and a *Latitude*: accounting the latitude towardes other Sciences : and the Longitude towards Action : that is, from the greateſt Generalitie , to the moſt particular Precept : The one

R r   giueth

giueth Rule howe farre one knowledge ought to intermeddle within the Prouince of another, which is the Rule they call Καθαυτο. The other giueth Rule, vnto what degree of particularitie, a knowledge fhould d fcend: which latter I finde paffed ouer in filence; being in my Iudgement, the more materiall. For certainely, there muft bee fomewhat left to practife ; but howe much is worthy the Enquirie : wee fee remote and fuperficiall Generalities, doe but offer Knowledge, to fcorne of practicali men : and are no more ayding to practife, than an *Ortelius* vniuerfall Mappe, is to direct the way betweene *London* and *Yorke*. The better fort of Rules, haue beene not vnfitly compared to glaffes of fteele vnpullifhed ; where you may fee the Images of thinges, but firft they muft bee filed : So *De Produ-* the Rules will helpe, if they bee laboured and pul-*ctione Ax-* lifhed by practife. But howe Chriftallyne they *iomatum.* may bee made at the firft, and howe farre forth they may be pullifhed afore-hand, is the queftion; the Enquirie whereof, feeemeth to me deficient.

There hath beene alfo laboured, and put in practife a *Methode*, which is not a lawfull *Methode*, but a *Methode* of *Impofture*; which is to deliuer knowledges in fuch manner, as men may fpeedily come to make a fhewe of Learning, who haue it not ; fuch was the trauaile of *Raymundus Lullius*, in making that *Art*, which beares his name ; not vnlike to fome Bookes of *Typocofmy*, which haue beene made fince ; beeing nothing but a Maffe

of

of words of all Arts; to giue men countenance, that thofe which vfe the tearmes; might bee thought to vnderftand the Art; which Colleftions are much like a Frippers or Brokers ſhoppe; that hath ends of euerie thing, but nothing of worth.

Nowe wee defcend to that part, which concerneth the ILLVSRATION OF TRADITION, comprehended in that Science, which wee call RHETORICKE, OR ART OF ELO-QVENCE; A Science excellent, and excellently well laboured. For although in true value, it is inferiour to Wifedome, as it is fayd by God to *Mofes*, when he difabled himfelfe, for want of this Facultie, *Aaron fhall bee thy Speaker, and thou fhalt bee to him as God :* Yet with people it is the more mightie ; For fo *Salomon* fayth : *Sapiens Corde appellabitur Prudens, fed dulcis Eloquio Maiora reperiet :* Signifying that profoundneffe of Wifedome, will helpe a Man to a Name or Admiration ; but that it is Eloquence, that preuayleth in an aftiue life ; And as to the labouring of it, the Emulation of *Ariftotle*, with the *Rhetoricians* of his time, and the experience of *Cicero*, hath made them in their Workes of *Rhetorickes*, exceede themfelues. Againe, the excellencie of Examples of *Eloquence*, in the Orations of *Demofthenes* and *Cicero*, added to the perfeftion of the Precepts of *Eloquence*, hath doubled the progrefsion in this Arte : And therefore, the Deficiences which I fhall note, will rather bee in fome Colleftions, which may as

Hand-

Hand-maydes attend the Art ; than in the Rules, or vſe of the Art it ſelfe.

Notwithſtanding, to ſtirre the Earth a little about the Rootes of this Science, as we haue done of the reſt ; The dutie and Office of *Rhetoricke* is, *To apply Reaſon to Imagination,* for the better moouing of the will ; For wee ſee *Reaſon* is diſturbed in the Adminiſtration thereof by three meanes ; by *Illaqueation,* or *Sophiſme,* which pertaines to *Logicke* ; by *Imagination* or *Impreſſion,* which pertaines to *Rhetoricke,* and by *Paſſion* or *Affectiun,* which pertaines to *Moralitie.* And as in Negotiation with others ; men are wrought by cunning, by Importunitie, and by vehemencie ; So in this Negotiation within our ſelues ; men are vndermined by *Inconſequences,* ſollicited and importuned, by *Impreſſions* or *Obuerſations*: and tranſported by *Paſſions*: Neither is the Nature of Man ſo vnfortunately built, as that thoſe Powers and Arts ſhould haue force to diſturbe Reaſon, and not to eſtabliſh and aduance it : For the end of *Logicke,* is to teach a fourme of Argument, to ſecure Reaſon, and not to entrappe it. The end of *Moralitie,* is to procure the Affections to obey Reaſon, and not to inuade it. The end of *Rhetoricke,* is to fill the Imagination to ſecond Reaſon, and not to oppreſſe it : for theſe abuſes of Arts come in, but *Ex obliquo,* for Caution.

And therfore it was great Iniuſtice in *Plate,* though ſpringing out of a iuſt hatred of the *Rhetoricians* of his time, to eſteeme of *Rhetoricke,* but as a voluptuarie Art,

Art, refembling it to Cookerie, that did marre whol-
fome Meates, and helpe vnwholefome by varietie of
fawces, to the pleaſure of the taſt. For wee ſee that
ſpeech is much more conuerſant in adorning that
which is good, than in colouring that which is euill:
for there is no man but ſpeaketh more honeſtly, than
he can doe or thinke ; And it was excellently noted
by *Thucidides* in *Cleon*, that becauſe he vſed to hold
on the bad ſide in Cauſes of eſtate; therefore hee
was euer inueying againſt Eloquence, and good
ſpeech; knowing that no man can ſpeake faire of
Courſes ſordide and baſe. And therefore as *Plato*
ſayd elegantly : *That vertue, if ſhee could be ſeen; would*
mooue great loue and affection : So ſeeing that ſhe cannot bee ſhewed to the *Sence*, by corporall ſhape, the
next degree is, to ſhewe her to the *Imagination* in
liuely repreſentation: for to ſhewe her to *Reaſon*, only in ſubtilitie of Argument, was a thing euer derided in *Chryſippus*, and many of the Stoykes, who
thought to thruſt vertue vppon men by ſharpe diſputations and Concluſions, which haue no Sympa-
thy with the will of Man.

Againe, if the affections in themſelues were
plyant and obedient to Reaſon, it were true,
there ſhoulde bee no great vſe of perſwaſions
and inſinuations to the will, more than of naked
propoſition and Proofes: but in regard of the continuall Mutinies and Seditions of the Affections :

Video meliora, Probeque; Deteriora ſequor ;

Reaſon would become Captiue and ſeruile, if

Eloquence

Eloquence of Perswasions, did not practife and winne the *Imagination*, from the *affections* part, and contract a Confederacie betweene the *Reafon* and *Imagination*, againft the *Affections* : For the Affections themfelues, carrie euer an appetite to good, as Reafon doth : The difference is, *That the Affection beholdeth meerely the prefent ; Reafon behouldeth the future, and fumme of time*. And therefore, the *Prefent*, filling the *Imagination* more ; *Reafon* is commonly vanquifhed; But after that force of *Eloquence* and *perfwafion*, hath made thinges *future*, and *remote*, appeare as *prefent*, than vppon the reuolt of the Imagination, Reafon preuayleth.

Wee conclude therefore, that *Rhetoricke* can bee no more charged, with the colouring of the worfe part, than *Logicke* with *Sophiftrie*, or Moralitie with Vice. For wee knowe the Doctrines of Contraries are the fame, though the vfe be oppofite: It appeareth alfo, that *Logicke* differeth from *Rhetoricke*, not onely as the *fift*, from the *pawme*, the one clofe, the other at large ; but much more in this, that *Logicke* handleth Reafon exacte, and in truth ; and *Rhetoricke* handleth it, as it is planted in popular opinions and Manners : And therefore *Ariftotle* doth wifely place *Rhetoricke*, as betweene *Logicke* on the one fide, and Morall or *Ciuile* Knowledge on the other, as participating of both: for the Proofes and Demonftrations of *Logicke*, are toward all men indifferent, and the fame: But the Proofes and perfwafions of *Rhetoricke*, ought to
differ

differ according to the Auditors,

Orpheus in Syluis, inter Delphinas Arion;

Which application, in perfection of *Idea*, ought to extend so farre: that if a Man should speake of the same thing to seuerall persons: he should speake to them all respectiuely and seuerall wayes: though this *Politique part of Eloquence in priuate Speech*, it is easie for the greatest Orators to want: whilest by the obseruing their well graced fourmes of speech, they leese the volubilitie of Application: and there- *De pruden-* fore, it shall not be amisse to recommend this to bet *tia sermo-* ter enquirie, not being curious, whether we place *nis priuati.* it heere, or in that part which concerneth Policie.

Nowe therefore will I descend to the deficiences, which (as I sayd) are but Attendances: and first, I doe not finde the Wisedome and diligence *Colores bo-* of *Aristotle* well poursued, who began to make *ni & mali,* a collection of *the popular signes and colours of good simplicis* *and euill, both simple and comparatiue,* which are *& compa-* as the *Sophismes* of *Rhetoricke,* (as I touched be- *rati.* fore.) For Example.

S O P H I S M A.

Quod laudatur, bonum: Quod vituperatur, malum.

R E D A R G V T I O.

Laudat vænales, qui vult extrudere merces.

Malum est, Malum est (inquit Emptor) sed cum re- *cesserit, tum gloriabitur.* The defects in the labour of *Aristotle* are three: One, that there be but a few of many: another, that their *Elenches* are not annex-

R r 4   ed;

ed ; and the third, that hee conceiued but a part of the vſe of them : for their vſe is not onely in probation, but much more in Impreſsion. For many fourmes are equall in *Signification*, which are differing in *Impreſsion:* as the difference is great in the piercing of that which is ſharpe , and that which is flat, though the ſtrength of the percuſsion be the ſame : for there is no man , but will be a little more rayſed by hearing it ſayd : *Your enemies will be glad of this,*

  Hoc Ithacus velit, & magno mercentur Atridæ,

Than by hearing it ſayd only, *This is euill for you.*

Secondly, I do reſume alſo, that which I mentioned before, touching *Prouiſion* or *Præparatorie ſtore,* for the Furniture of ſpeech, and readineſſe of Inuention ; which appeareth to be of two ſorts ; The one in reſemblance to a ſhoppe of peeces vnmade vp; the other to a ſhopp of thinges ready made vp, both to be applyed to that which is frequent, and moſt in requeſt ; The former of theſe I will call *Antithera,* & the latter *Formulæ.*

Antitheta are *Theſes* argued, *pro & contra,* wherin men may be more large & laborious ; but (in ſuch as are able to doe it) to auoyd prolixity of entry, I wiſh the ſeedes of the ſeuerall arguments to be caſt vp into ſome briefe and acute ſentences : not to bee cyted : but to bee as Skaynes or Bottomes of thread, to bee vnwinded at large , when they come to be vſed : ſupplying authorities, and Examples by reference.

Antithe-
ſa rerum.

Pro verbis legis,
Non est interpretatio, sed diuinatio, quæ recedit a littera,
Cum receditur a littera Iudex transit in legislatorem,
Pro sententia Legis.
Ex omnibus verbis est Elucendus sensus, qui interpretatur
singula:

Formulæ are but decent and apt paſſages or conueyances of ſpeeche, which may ſerue indifferently for
differing ſubiects, as of *Preface, Concluſion, Digreſſion,*
Tranſition, Excuſation, &c. For as in buildings there is
great pleaſure and vſe in the well caſting of the ſtaire
caſes, entryes, doores, windowes, and the like, ſo in
ſpeeche, the conueyances and paſſages are of ſpeciall
ornament and effect.

A concluſion in a Deliberatiue.
So may we redeeme the faults paſſed & preuent the inconue
niences future.

There remayn two Appendices touching the tradition of knowledge, The one *Criticall*, The other *Pe-*
danticall. For all knowledge is eyther deliuered by
Teachers, or attayned by mens proper endeuors:
And therefore as the principall part of Tradition of
knowledge concerneth chiefly in *writing of Books*; So
the Relatiue part thereof concerneth *reading of Bookes*
Wherunto appertayn incidently theſe conſideratiõs.
The firſt is cõcerning the true Correction & editiõ of
Authors, wherin neuertheleſſe raſh diligẽce hath don
gret preiudice. For theſe *Critiques* haue ofte preſumed
that that which they vnderſtand not, is falſe ſet down;
As the Prieſt, that where he found it written of *S. Paul*
Demiſſus eſt per ſportam, mẽded his book, and made it *De-*
Sſ   *miſſus*

miſſus eſt per portam becauſe, *Spcrta* was an hard word, and out of his reading; and ſurely their errors, though they be not ſo palpable and ridiculous, yet are of the ſame kind. And therefore as it hath beene wiſely noted, the moſt corrected copies are cōmonly the leaſt correct.

The ſecond is concerning the expoſition and explication of Authors, which reſteth in Annotacions and Cōmentaryes, wherin it is ouer vſual to blaunch the obſcure places, and diſcourſe vpon the playne.

The third is concerning the times, which in many caſes giue great light to true Interpretations.

The fourth is concerning ſome briefe Cenſure and iudgement of the Authors, that men therby may make ſome election vnto themſelues, what Bookes to reade:

And the fiſt is concerning the Syntax and diſpoſition of ſtudies, that men may know in what order or purſuite to reade.

For P E D A N T I C A L L knowledge, it contayneth that differēce of *Tradition* which is proper for youth: Whereunto appertaine diuers conſiderations of great fruit.

As firſt the tyming and ſeaſoning of knowledges, as with what to initiate them, and from what for a time to refraine them:

Secondly, the conſideration where to begin with the eaſieſt, and ſo proceede to the more difficult, And in what courſes to preſſe the more difficulte and then to turne them to the more eaſie : for it is one

Me

Methode to practife fwimming with bladders, and a-nother to practife dauncing with heauy fhooes.

A third is the application of learning according vnto the propriety of the wittes; for there is no defect in the faculties intellectuall, but feemeth to haue a proper Cure contayned in fome ftudies; As for example, If a Child be Bird-witted, that is, hath not the facultie of attention, the Mathematiques giueth a re-medy thereunto; for in them, if the witte be caught away but a moment, one is new to begin. And as fci-ences haue a propriety towards faculties for Cure and helpe; So faculties or powers haue a Simpathy to-wards Sciences for excellency or fpeedy profiting: And therfore it is an enquiry of greate wifedom what kinds of wits and Natures are moft apt and proper for what fciences.

Fourthly the ordering of exercifes is matter of great confequence to hurt or helpe; For as is well ob-ferued by *Cicero*, men in exercifing their faculties if they be not wel aduifed doe exercife their faultes & get ill hàbits aswell as good; fo as there is a greate iudgement to be had in the continuance and inter-miffion of Exercifes. It were to longe to particularize a number of other confideratiõs of this nature, things but of meane appearance, but of fingular efficacy. For as the wronging or cherifhing of feeds or young plants, is that, that is moft important to their thriuing And as it was noted, that the firft fix kings being in trueth as Tutors of the State of Rome in the infancy thereof, was the principal caufe of the immenfe great-
S f 2   neffe

neſſe of that ſtate which followed. So the culture and manurance of Minds in youth, hath ſuch a forcible (though vnſeen) operacion, as hardly any length of time or contention of labour can counteruaile it afterwards. And it is not amiſſe to obſerue alſo, how ſmall and meane faculties gotten by Education, yet when they fall into greate men or great matters, doe work great and important effects : whereof we ſee a notable example in *Tacitus* of two Stage-plaiers, *Percennius* and *Vibulenus*, who by their facultie of playing, put the *Pannonian* armies into an extreame tumulte and combuſtion. For there ariſing a mutinie amongſt them, vpon the death of *Auguſtus Cæſar*, *Blæſus* the lieuetenant had committed ſome of the Mutiners which were ſuddenly reſcued: whereupon *Vibulenus* got to be heard ſpeake, which he did in this manner, *Theſe poore innocent wretches appointed to cruell death, you haue reſtored to behould the light. But who ſhall reſtore my brother to me, or life vnto my brother? that was ſent hither in meſſage from the legions of Germany, to treat of the common Cauſe, and he hath murdered him this laſt night by ſome of his fencers & ruffians, that he hath about him for his executioners vpon Souldiours : Anſwer Blæſus, what is done with his body: The mortalleſt Enemies doe not deny buriall: when I haue performed my laſt duties to the Corpes with kiſſes, with teares, command me to be ſlaine beſides him, ſo that theſe my fellowes for our good meaning, and our true hearts to the Legions may haue leaue to bury vs.* With which ſpeeche he put the army into an infinite fury and vprore, whereas truth was he had no
bro

brother, neyther was there any such matter, but hee plaide it meerely as if he had beene vpon the stage.

But to returne, we are now come to a period of RATIONALL KNOVVLEDGES, wherein if I haue made the *diuisions* other than those that are receiued, yet would I not be thought to disallow all those diui sions, which I doe not vse. For there is a double ne cessity imposed vpon me of altering the diuisions. The one because it differeth in end and purpose, to sorte together those things which are next in Nature, and those things which are next in vse. For if a secretary of Estate, should sort his papers, it is like in his study, or generall Cabinet, he would sort together things of a Nature, as Treaties, Instructions, &c. But in his Boxes, or particular Cabinet, hee would sort togi ther those that he were like to vse together, though of seuerall Natures : So in this generall Cabynet of knowledge, it was necessary for me to follow the di uisions of the Nature of things, whereas if my selfe had beene to handle any particular knowledge, I would haue respected the *Diuisions fitest for vse.* The other, because the bringing in of the *Deficiences* did by Consequence alter the *Partitions* of the rest . For let the knowledge extant (for demonstration sake) be 15. Let the knowledge with the Deficiences be 20. the parts of 15 are not the parts of 20, for the parts of 15, are 3.and 5 the parts of 20. are 2.4.5.and 10. So as these things are witbout Contradiction, and could not otherwise be,

We

WE proceed now to that knowledge which cõ
sidereth of the APPETITE and WILL
OF Man, whereof *Salomon fayth ᴧnte omnia
fili cuſtodi cor tuum,nam inde procedunt actio-
res vit.e.*In the handling of this ſcience, thoſe which
haue written ſeeme to me to haue done as if a man
that profeſſed to teach to write did only exhibit faire
copies of *Alphabets,*& letters ioyned, without geuing
any precepts or directiõs,for the cariage of the hãd &
framing of the letters.So haue they made good& fair
Exemplars & coppies,carieng the draughts and pour
traiturs of *Good.Vertue,Duety,Felicity,*propoũding thẽ
well deſcribed as the true obiects and ſcopes of mãs
wil and deſires:But how to attain theſe excellẽt marks,
and how to frame and ſubdue the will of man to become true and conformable to theſe purſuites,they
paſſe it ouer altogether,or ſlightly and vnprofitably
For it is not the diſputing.That morall vertues are in
the Minde of man by habite & not by nature: or the
diſtinguiſhing.That generous ſpirites are wonne by
doctrines and perſwaſions,and the vulgar ſort by reward & puniſhment, and the like ſcattered glances
and touches that can excuſe the abſence of this
parte.

The reaſon of this omiſſion I ſuppoſe to be that
hidden Rocke wherevppon both this and many other
barques of knowledge haue beene caſt away, which
is ,that men haue diſpiſed to be conuerſant in ordinary and common matters , the iudicious direction
whereof neuertheleſſe is the wiſeſt doctrine: (for life
con-

consisteth not in nouelties nor subtilities) but contrariwise they haue compounded Sciences chiefly of a certaine resplendent or lustrous masse of matter chosen to giue glory either to the subtillity of disputacions or to the eloquence of discourses. But *Seneca,* giueth an excellent check to eloquence *Nocet illis eloquentia,quibus non rerum cupiditatem facit sed sui* ,doctrines should be such as should make men in loue with the Lesson ,and not with the Teacher, being directed to the Auditors benefite,and not to the Authors commendation:And therefore those are of the right kinde which may be concluded as *Demosthe.nes* concludes his counsell *Quæ si feceritis non Oratorem duntaxat in præsentia laudabitis sed vosmetipsos etiã nõ ita multo post statu rerum vestrarum meliore.*

Neyther needed men of so excellent parts to haue despaired of a Fortune,(which the Poet *Virgill* promised himselfe,and indeed obtained) who got as much glory of eloquence, wit, and learning in the expressing of the obseruacions of husbandry , as of the heroicall acts of *Æneas.*

Nec sum animi dubius verbis ea vincere magnum,
Quam sit & angustis his addere rebus honerem.

And surely if the purpose be in good earnest not to write at leasure that which mē may read at leasure,but really to instruct and suborne Action and actiue life, these Georgickes of the mind concerning the husbãdry & tillage therof,are no lesse worthy thē the heroical descriptiõs of *vertue,duty,& felicity* wherfore the maine &primitiue diuision of *Morall* knowledge seemeth

meth to be into the EXEMPLAR or PLATFORMB of GOOD, and the REGIMENT or CVLTVRE OF THE MIND; The one deſcribing the nature of Good the other preſcribing rules how to ſubdue, apply and accomodate the will of man therevnto.

The Doctrine touching the PLATFORME or NATVRE of GOOD conſidereth it either SIMPLE or COMPARED either the kindes of Good or the degrees of Good: In the later whereof thoſe infinite diſputations, which were touching the ſupreme degree thereof, which they terme Felicity, Beatitude, or the higheſt Good, the doctrines concerning which were as the heathen Diuinity, are by the chriſtian faith diſcharged. And as *Ariſtotle* ſaith, *That yong men may bee happy, but not otherwiſe, but by Hope*; So we muſt all acknowledge our Minority, and embrace the felicity, which is by hope of the future world.

Freed therefore, and deliuered from this doctrine of the Philoſophers heauen, whereby they fayned an higher eleuation of Mans Nature, then was; For we ſee in what an height of ſtile *Seneca* writeth, *Vere Magnum, habere fragilitatem hominis, ſecuritatem Dei.*) We may with more ſobriety and truerh receiue the reſt of their Enquiries, and labors? Wherein for the *Nature of Good Poſitiue, or ſimple*, they haue ſet it downe excellently, in deſcribing the fourmes of *Vertue* and *Duty*, with their ſituations and poſtures, in diſtributing them into their kinds, parts, Prouinces, Actions, and Adminiſtrations, and the like; Nay furder, they haue commended them to Mans Nature, and ſpirite

with

with greate quicknesse of Argumente, and beauty of persuasions, yea , and fortified and entrenched them (as much as discourse can doe)againft corrupt and popular opinions. Againe , *for the degrees, and Comparatiue Nature of Good* , they haue also excellentlye handled it in their triplicity of *Good*; in the comparisons betweene a Contemplatiue and an actiue life,in the diftinction between vertue with reluctation,and vertue secured; n their encounters between honefty and profit, in theyr ballancing of vertue with vertue, and the like; fo as this parte deferueth to bee reported for excellentlye laboured.

Notwithftanding, if before they had commen to the popular and receiued Nocions of vertue and vice,pleafure and payne,and the reft, they had ftayed a little longer vpon the Enquirye , concerning the Rootes of Good and euill, and the Strings of those Rootes,they had giuen in my opinion,a great light to that which followed; and fpeciallye if they had coufulted with Nature,they had made their doctrins leffe prolixe,and more profound; which beeing by them in part omitted,and in part handled with much Confufion,we will endeauour to refume, and open in a more cleare Manner.

There is fourmed in euery thing a double Nature of Good;the one,as euery thing is, a Totall or fubftantiue in it felfe; the other,as it is a parte or Member of a greater Bodye ; whereof the

T t   later

the later is in degree the greater, and the worthier, becaufe it tendeth to the conferuation of a more ge= nerall fourme. Therefore we fee, the Iron in particu- ler fimpathye mooueth to the Loadftone; But yet if it exceede a cettayne quantity, it forfaketh the affecti- on to the *Loadftone* and like a good patriot mooueth to the *Earth* which is the Region and Countrye of Maffie Bodyes; fo may we goe forward, and fee that *Water* and *Maffie bodyes* moue to the *Center of the earth* But rather thē to fuffer a diuuifiō in the cōtinuāce of Nature they wil mooue vpwards from the Center of the Earth: forfaking their dutye to the *Earth* in re= gard of their duty to the *World.* This double nature of Good & the com- paratiue thereof is much more engrauē vpon Man, if he degenerate not: ʌnto whō the Cōferuation of duty to the publique ougʰt to be much mɔre pecious then the Conferuation of life and being: according *to* that Memorable fpeache of *Pompeius Magnus* .when being in commiffion of puruciance for a famine at Rome, and being diffua- ded with great vehemency and inftance by his frinds about him that he fhould uot hazard himfelfe to Sea in an exreemity of weather he fayd only to them: *Neceffe eft vt eam, non vt viuam:* But it may be truly affirmed that there was neuer any phyɪofophy, Reli- gion or other difcipline, which dıd fo playnly and highly exalt the good which is *Communicatiue* and depreffe the good which is priuate and partriculer as the Holy faith: well declaring that it was the fame God, that gaue the Chriftian Law to men, who gaue

thofe

thofe Lawes of nature, to inaminate Creatures that we fpake of before ; for we reade that the elected Saints of God haue wifhed themfelues Anathemized, and razed out of the Booke of life, in an extafie of Charity, and infinite feeling of *Communion*.

This being fet downe and ftrongly planted doth iudge and determine moft of the Controuerfies wherein *Morall Philofophie* is Conuerfant ; For firft it decideth the queftion touching the preferment of the Contemplatiue or actiue life, and decideth it againft *Ariftotle*; For all the reafós which he bringeth for the Contemplatiue, are priuate, and refpecting the pleafure and dignity of a mans felfe. (in which refpects no queftion the contemplatiue life hath the preemynence;) not much vnlike to that Comparifon, which *Pythagoras* made for the gracing and Magnifying of Philofophy, and Contemplacion who being afked what he was, anfwered : *That if Hiero were euer at the Olimpian games, he knew the Manner, that fome came to try their fortune for the prizes, and fome cam as Merchants to vtter their commodities, and fome came to make good cheere, and meete their friends, and fom came to looke on, & that he was one af them that came to look on.* But men muft know, that in this Theater of Mans life, it is referued onely for God and Angels to be loo kers on, Neither could the like queftion euer haue beene receiued in the Church, notwithftanding their (*Pretiofa in oculis Domini mors fanctorum eius*) by which place they would exalt their Ciuile death, and regular profeffions, but vpon this defence, that the Monafticall

naſtical life is not ſimple Contemplatiue, but perfor-
meth the duty either of inceſſant prayers and ſuppli-
cations which hath been truly eſteemed as an office
in the church, or els of writing or in taking inſtruc-
tions for writing concerning the law of God as *Moſes* did, when he abode ſo long in the Mount. And ſo
wee ſee Henoch the .7. from Adam who was the firſt
Contemplatiue & walked with God, yet did alſo en-
dow the Church with prophefy which *Sainte Iude*
citeth. But for contemplation which ſhould be finiſhed in it ſelfe without caſting beames vpon ſociety,
aſſuredly diuinity knoweth it not.

It decideth alſo the controuerſyes betweene *Zeno*
and *Socrates*, and theyr ſchooles and ſucceſſions on
the one ſide, who placed felicity in vertue ſimply or
attended: the actions and exerciſes wherof do chiefly
imbrace and concerne ſociety; & on the other ſide,
the *Cirenaiques* & *Epicureans*, who placed it in pleaſure
and made vertue, (as it is vſed in ſome comedyes of
Errors, wherein the Miſtres and the Maide change ha
bits) to be but as a ſeruāt, without which, pleaſure cānot be ſerued and attended, and the reformed ſchoole
of the Epicureās, which placed it in ſerenity of mind
and freedome from perturbation: as if they woulde
haue depoſed *Iupiter* againe, and reſtored *Saturne*, and
the firſt age, when there was no ſummer nor winter,
ſpring nor Autumne, but al after one ayre and ſeaſon
And *Herillus*, which placed felicity in extinguiſhmēt
of the diſputes of the mind, making no fixed Nature
of Good and euill, eſteeming things according to the

cleer-

cleernes of the defires, or the reluctation: which opini
on was reuiued in the herefy of the Anabaptifts, mea
furing things according to the motions of the fpirit,
and the conftancy or wauering of beleefe, all which
are manifeft to tend to priuate repofe & cōtentment,
and not to poynt of fociety.

It cenfureth alfo the philofophy of *Epictetus* which
prefuppofeth that felicity muft bee placed in thofe
things which are in our power, leaft we be lyable to
fortune & difturbance: as if it were not a thing much
more happy to faile in good and vertuous ends for
the publicke, then to obtayne all that wee can wifh to
our felues in our proper fortune: as *Confaluo* fayd to
his fouldiers, fhewing them *Naples* and protefting, he
had rather dy one foote forwards, then to haue his
life fecured for long, by one foote of retrayt: Whereunto the wifedome of that heauenly Leader hath figned, who hath affirmed that *A good Confcience is a
continuall Feafte*, fhewing plainly that the confcience
of good intencions howfoeuer fucceeding, is a more
continuall ioy to nature, then all the prouifion which
can be made for fecurity and repofe.

It cenfureth likewife that abufe of Philofophy, which
grew generall about the time of *Epictetus*, in conuer-
ting it into an occupation or profeffion: as if the purpofe had bene, not to refift and extinguifh perturbations, but to fly and auoide the caufes of them, & to
fhape a particular kind and courfe of life to that end,
introducing fuch an health of mind, as was that heaith
of body, of which *Ariftotle* fpeaketh of *Herodicus*, who
did

did nothing nothing all kis life long, but intend his health, whereas if men refer themfelues to dutyes of Society; as that health of Body is beft, which is ableft to endure all alterations and extremityes, So likewife that health of Mind is moft proper , which can goe through the greateft temptations and perturbations. So as *Diogenes* opinion is to be accepted, who Commended not them which abfteyned, but them which fuftayned, and could refraine their Mind in *Precipitio*, and could giue vnto the mind (as is vfed in horfman-fhip) the fhorteft ftop or turne.

Laftly it cenfureth the Tendernefle and want of application in fome of the moft auncient and reuerend Philofophers and Philofophicall men, that did retyre too eafily from Ciuile bufinefle, for auoyding of Indignities & perturbations, whereas the refolution of men truly Moral, ought to be fuch, as thefame *Confaluo* fayd, the honor of a fouldior fhould be *F te- la Craffiore*, and not fo fine, as that euery thing fhould catch in it, and endanger it.

To refume *priuate* or *particular good*; it falleth into the diuifiõ of *Good Actiue & Paffiue*; For this differéce of *Good*, (not vnlike to that which amongft the Romãs was expreffed in the familiar or houfhold terms of *Promus*, and *Condus*,) is formed alfo in all things, & is beft difclofed in the two feuerall Appetites in creatures; the one to preferue or continue themfelues, & the other to dilate or Multiply themfelues; whereof the later feemeth to be the worthyer; For in Nature the heauens, which are the more worthy, are the *A*

gen

gent, and the earth , which is the leſſe woorthye is
the Patient . In the pleaſures of liuing creatures ,
that of generation is greater then that of foode.
In diuine Doctrine , *Beatius eſt dare quam accipere :*
And in life there is no mans ſpirit ſo ſoft but eſte-
meth the effecting of ſomwhat that he hath fixed in
his deſire more then ſenſuality which priority of the
Actiue Good, is much vpheld by the Conſideration
of our eſtate to be mortall & expoſed to fortune: for,
if wee mought haue a perpetuity and Certainty in
our pleaſures the *State* of them would advance their
price. But when we ſee it is but *Magni æſtimamus Mori*
tardius and *Negligeris de craſtino. Neſcis Partum diei*
it maketh vs to deſire to haue ſomwhat ſecured and
exempted from Time, which are onelye our deedes
and works, as it is ſayd *Opera eorum ſequuntur eos.* The
preheminence likewiſe of this actiue good is vpheld
by the affection which is naturall in man towardes
variety and proceeding which in the pleaſures of the
ſence which is the principal part of *Paſſiue* good)can
haue no great latitude. *Cogita quamdiu eadem feceris Ci*
bus, Somnus Ludus per hunc Circulū curritur, mori velle
nō tantū ſortis aut miſer aut prudens ſed etiā faſtidioſus po-
teſt. But in enterpriſes, purſutes & purpoſes of life ther
is much variety, wherof men are ſēſible with pleaſure
in theyr inceptions, progreſſions, recoyls, reintegrati-
ons, approches and atteynings to their ends. So as it
was wel ſaid: *Vita ſiue propoſito languida & vaga eſt.* Neither hath this Actiue good and Identity with the
good of Society though, in ſome caſe, it hath an incidence

cidence into it: For although it do many times bring forth Acts of *Beneficēce* yet it is with a respect priuate to a mās own power, glory, amplificatiō, cōtinuāce: as appeareth plainly when it findeth a contrary Sùbiect For that Gygātine state of mind which posseseth the trowblers of the world, such as was *Luoius Sylla* and infinit other in smaller model who would haue all mē happy or vnhappy as they were their friends or Enimies, and would giue fourm to the world according to their owne humors (which is the true *Theomachy* pretendeth and aspireth to Actiue good, though it recedeth furthest from good of Society which wee haue determined to be the greater.

To resume *Passiue Good* it receiueth a subdiuision of *Conseruatiue* and *Perfectiue.* For let vs take a brief Review of that which we haue said, we haue spoken first of the Good of Society the intention whereof embraceth the Fourm of Humaine Nature, whereof we are members & Portions: and not our owne proper and Indiuidual fourme: we haue spoken of Actiue good and supposed it as a part of Priuate and particular good. And rightly: For there is impressed vppon all things a triple desire or appetite proceeding from loue to themselues, one of *preseruing and contynuing* theyr form, another of *Aduancing* and *Perfitting* their fourm and a third of *Multiplying* and extending their fourme vpon other things: whereof the multiplying or signature of it vpon other things, is that which we handled by the name of Actiue good. So as there remayneth the conseruing of it and parfiting or raising

of

of it:which later is the higheſt degree of Paſſiue good
For to preſerue in ſtate is the leſſe, to preſerue with
aduancement is the greater. So in man

Igneus eſt ollis vigor,& cæleſtis origo. His approach
or Aſſumptiõ to diuine or Angellicall Nature,is the
perfeƈtion of his forme,The error or falſe Imitatiõ of
which good is|that whichis the tẽpeſt of bumane life
whileman vpõ the inſtinƈt of an aduãcement *Formaˡ.*
and *Eſſential* is carried to ſeek an aduancement *Locall*
For as thoſe which are ſick,& finde no remedy, doe
tumble vp and downe and chaunge place, as if by a
Remoue Locall, they could obtayne a Remooue In-
ternall:So is it with men in ambition,when ſayling of
the meane to exalt their *Nature*, they are in a perpetuall eſtuation to exalte theyr *Place*. So then *paſſiue
Good*, is, as was ſayde, eyther *Conſeruatiue* or *Per-
feƈtiue.*

To reſume the good of *conſeruation* or *Comforte*,
which conſiſteth *in the fruicion of that which is agreeable to our Natures*,it ſeemeth to be the moſt pure and
Naturall of pleaſures, but yet the ſofteſt and the low-
eſt.And this alſo receiueth ã differẽce,which hath neither beene well iudged of, nor well inquired· For
the good of fruition or contentment,is placed eyther
in the *Sincereneſſe of the fruition*,or in the *quickneſſe &
vigor of it*,the one ſuperinduced by the *Æquality*,the o
ther by *Vicicitude*:the one hauing leſſe mixture of *Euils*
the other more impreſſiõ of *Good*.Whether of theſe,is
the greter good,is a queſtiõ cõtrouerted,but whether

V v   mans

maus nature m ay not be capable of both, is a question not inquired.

The former queſtion heing debated between *St-
crates*, and a *Sophiſt*, *Socrates* placing felicity in an e-
quall and conſtant peace of mind;and the Sophiſt in
much deſiring,and much enioying:they fell from Argument to ill words: The Sophiſt ſaying that *Socrates*
felicity,was the felicity of a block or ſtone,and *Socrates* ſaying that th e *Sophiſts* felicity, was the felicity of
one that had the itch, who did nothing but itche and
ſkratch.And both theſe opinions, do not want their
ſupports.For the opinion of *Socrates* is much vpheld
by the generall conſent,euen of the Epicures themſelues,that vertue beareth a great part in felicity:and
if ſo,certain it is, that vertue hath more vſe in clee-
ring perturbations, then in compaſſing deſires. The
Sophiſts opinion is mnch fauoured,by the Aſſertion
we laſt ſpake of,that *good of Aduancement* , is greater
then *good of ſimple Preſeruation* : becauſe, euery obtayning a deſire,hath a ſhew of aduancement,as mocion
though in a Circle,hath a ſhew of progreſſion.

But the ſecond queſtion, decided the true waye,
maketh the former ſuperfluous.For, can it be doubted,but that there are ſome, who take more pleaſure
in enioying pleaſures,then ſome other; and yet neuertheleſſe,are leſſe troubled with the loſſe or leauing of them?So as this ſame; *Non vti,vt non appetas :
Non appetere,vt non metuas,ſunt animi puſilli & diffiden-
tis.* And it ſeemeth to me, that moſt of the doctrines
of the Philoſophers are more fearefull and cautionary
then

then the Nature of things requireth. So haue they en-
creafed the feare of death, in offering to cure it. For,
when they would haue a mans whole life, to be but a
difcipline or preparation to dye : they muft needes
make men thinke, that it is a terrible Enemy, againft
whom there is no end of preparing. Better faith the
Poet,

Quæ finem vitæ extremum inter Munera ponat
Naturæ: So haue
they fought to make mens minds to vniforme and
harmonicall, by not breaking them fufficiently to cō-
trary Motions: the reafon whereof, I fuppofe to be,
becaufe they themfelues were men dedicated, to a pri
uate, free, and vnapplied courfe of life. For, as we fee,
vpon the lute, or like Inftrument, a *Ground*, though it
be fweet, and haue fhew of many changes, yet brea-
keth not the hand to fuch ftrange and hard ftoppes
and paffages, as a *Set fong*, or *Voluntary*: much after the
fame Manner was the diuerfity betweene a Philofophicall and a ciuile life. And therefore men are to I-
mitate the wifedome of Iewellers, who, if there be a
graine, or a cloude, or an ife which may be ground
forth, without taking to much of the ftone, they help
it: but, if it fhould leffen and abate the ftone to much
they will not meddle with it: So ought men, fo to procure *Serenity*, as they deftroy not magnanimity.

Hauing therefore deduced the *Good of Man, which
is priuate & particular*, as far as feemeth fit : wee will
now returne to that *Good of man, which refpecteth and be*

beholdeth Society which we may terme Duty; bi-
caufe the term of duty is more propper to a minde
well framed & difpofed towards others, as the terme
of vertue is applyed to a mind well formed & cōpo-
fed in it felfe, though neither can a man vnderftand
vertue without fome relation to Society, nor duety
without an inwarde difpofition, This part may feem
at firft to pertaine to Science Ciuile and Politicke:
but not if it be wel obferued, For it concerneth tho
Regimēt & gouernment of euery man, over himfelf,
& not ouer others. And as in architectur, the directiō
framing the poftes beames and other parts of building is not the fame with the maner of ioyning them
aud erecting the building: And in mechanicalls, the
direction how to frame an Inftrument or Engyne,
is not the fame with the manner of fetting it on
woorke and imploying it: and yet neuertheleffe in
expreffing of the one, you incidently expreffe the
Aptneffe towardes the other: So the doctrine of
Coniugation of men in Socyety, differereth from
that of their conformity therevnto.

This part of Duty is fudiuided into two parts: the
common duty of euery man, as a Man or member of
a State: The other the refpectiue or fpeciall duty of
euery man in his profeffion vocation and place:
The firft of theas, is extāt & wel laboured as hathbeen
faid. The fecōd like wife I may report rather difperfed
thē deficiēt: which maner of difperfed writing in this
kind of Argumēt, I acknowledge to be beft. For who
cā take vpō him to write of the proper duty, vertue cha
lenge and

and right, of euery feuerall vocation profeffion, and
place. For although fometimes a Looker on may fee
more then a gamefter and there be a Prouerb more
arrogant then found *That the vale beft difcouereth the
hill*: yet there is fmall doubt but that men can write
beft and moft really & materialy in their owne profeffions: & that the writing of fpeculatiue men of
Actiue Matter, for the moft part doth feeme to men
of Experience as Phormioes Argument of the warrs
feemed to Hannibal, to be but dreames and dotage.
Onely there is one vice which accompanieth them,
that write in their own profeffions that they magnify
thē in exceffe, But generally it were to be wifhed, (as
that which wold make learning indeed folide & fruit
ful) that Actiue men woold or could become writers
    In which kind I cannot but mencion *Honoris eaufa*
your Maiefties exellent book touching the duty of
a king: a woorke ritchlye compounded of *Diuinity
Morality and Policy*, with great afperfion of all other
artes: & being in myne opinion one of the mofte
found & healthful writings that I haue read: not dift-
empered in the heat of inuention nor in the Could-
nes of negligence: not fick of Dufineffe as thofe are
who leefe themfelues in their order; nor of Convul-
fions as thofe which Crampe in matters impertinent: not fauoring of perfumes & paintings as thofe
doe who feek to pleafe the Reader more then Nature beareth, and chiefelye wel difpofed in the fpirits
thereof, beeing agreeable to truth, and apt for actions: and farre remooued from that Naturall
                                                    infir-

infirmity, whereunto I noted thofe, that write in their own profeffions to be fubiect, which is, that they exalt it aboue meafure. For your Maiefty hath truly defcribed, not a king of Affyria, or Perfia, in their extern glory: but a *Mofes*, or a *Dauid*, Paftors of their people. Neither can I euer leefe out of my remembraunce, what I heard your Maiefty, in the fame facred fpirite of Gouernment, deliuer, in a great caufe of Iudicature which was: *That Kings ruled by theyr lawes, as God did by the lawes of Nature, and ought as rarely to put in vfe theyr fupreme Prerogatiue, as God doth his power of working Miracles*. Aud yet notwithftandiug, in your bookof a free Monarchy, you do well giue men to vnderftand, that you know the plenitude of the power and right of a King, as well as the Circle of his office and duty. Thus haue I prefumed to alledge this excellent writing of your Maiefty, as a prime or eminent example of *Tractates*, concerning fpeciall & refpectiue dutyes: wherin I fhould haue faid as much, if it had beene written a thoufand yeares fince: Neither am I mooued with certain Courtly decencyes, which efteeme it flattery to prayfe in prefence. No, it is flattery to prayfe in abfence: that is, when eyther the vertue is abfent, or the occafion is abfent: and fo the prayfe is not Naturall, but forced, either in truth, or in time. But let *Cicero be* read in his *Oration pro Marcello*, which is nothing but an excellent Table of *Cæfars* vertue, and made *to his face*, befides the example of many other excellent perfons, wifer a great deale then fuch obferuers : and we will neuer doubt, vpon a full occafion, to giue iuft

praifes

prayſes to preſent or abſent.

But to return, there belongeth further, to the handling of this partie touching the duties of profeſſions and vocations a *Relatiue or oppoſite* touching the fraudes cautels, impoſtures, & vices of euery profeſſion, which hath been likewiſe handled But howe? rather in a Satyre & Cinicaly, then ſeriouſly & wiſely for men haue rather ſought by wit to deride and traduce much of that which is good in profeſſions then with Iudgement to diſcouer and ſeuer that which is corrupt. For as Salomon ſaith. He that cometh to ſeeke after knowledg with a mind to ſcorne and cenſure, ſhalbe ſure to finde matter for his humor but no matter for his Inſtruction. *Quærenti deriſori Scientiam, ipſa ſe abſcondit: ſed Studioſo fit obuiam.* But the managing of this argument with integrity & Truthe, which I note as deficient, ſeemeth to me to be one of the beſt fortifications for honeſty and vertue that can be planted. For, as the fable goeth of the *Baſiliſk*, that if he ſee you firſt you die for it: but if you ſee him firſt, he dieth So is it with deceits and euill arts: which if they be firſt eſpied they leeſe their life, but if they preuent they indanger. So that we are much beholden to *Macciauell* & others that write what men doe and not what they ought to do. For it is not poſſible to ioyn ſerpentine wiſedom with the Columbine Innocency, except men know exactly all the conditions of the *Serpent*: his baſeneſſe and going vpon his bellye, his volubility and lubricity his enuy and ſtinge, and the reſt, that is al fourmes and Natures of euill. For
with

without this vertue lyeth open and vnfenced. Nay an honeſt man can doe no good vppon thoſe that are wicked to reclaime them, without the helpe of the knowledge of euil. For mē of corrupted minds preſuppoſe, that honeſty groweth out of Simplicitye of manners, and beleuing of Preachers, ſchoolmaſters, and Mens exterior language. So as, except you can make them perceiue, that you know the vtmoſt reaches of theyre owne corrupt opinions, they deſpiſe all moralitye. *Non recipit ſtultus verba prudentiæ, niſi ea dixeris, quæ verſantur in Corde eius.*

Vnto this part touching *Reſpectiue duty*, doth alſo appertayne the dutyes betweene huſband and wife, parent and childe, Maſter and Seruant: So likewiſe, the lawes of friendſhip and Gratitude, the ciuile bond of Companyes, Colledges, and Politike bodies, of neighbourhood, and all other proportionate duties: not as they ar parts of Gouernment and Society, but as to the framing of the minde of particular perſons.

The knowledge concerning *good reſpecting Society* doth handle it alſo not *ſimply* alone but *Comparatiuely* whereunto belongeth the weighing of duties, betwen perſon and perſon, Caſe and Caſe, particular & publike: As we ſee in the proceeding of *Lucius Brutus*, againſt his own Sons, which was ſo much extolled; yet what was ſayd?

Infelix, vtcunque ferent ea fata Minores.

So the caſe was doubtfull, and had opinion
on

on both fides: Againe we fee, when *M .Brutus* and
Caffius inuited to a fupper certaine, whofe opinions
they meant to feele,whether they were fit to be made
their Affociates , and caft foorth the queftion touching the *Killing of a Tyrãt* being an vfurper they wer
deuided in opinion,fome holding,that Seruitude was
the Extreame of Euils;and others, that Tyranny was
better,then a Ciuile war:and a number of the like cafes there are,of cõparatiue duty. Amõgft which, that
of all others,is the moft frequent,where the queftion
is of a great deale of good to enfue of a fmall Iniuft-
ice. Which *Iafon* of *Theffalia* determined againft the
truth; *Aliqua funt iniuftè facienda, vt multa iufte fieri
poffint.*But the reply is good; *Authorem præfentis Iufti-
titiæ habes fponforem futuræ non habes;* Men muft pur-
fue things which are iuft in prefente , and leaue the
future to the diuine prouidence:So then we paffe on
from this generall part touching the Exemplar and
defcription of Good.

  Now therefore,that we haue fpoken of this fruite
of life,it remaineth to fpeake of the Hufbandry that
belongeth thereunto,without which part, the former *De cul-*
feemeth to be no better then a faire Image,or *ftatua,* *tura ,*
which is beautifull to contemplate,but is without life *Animi.*
and mocion:whereunto *Ariftotle* himfelfe fubfcribeth
in thefe words : *Neceffe eft fcilicet de virtute dicere, &
quid fit,& ex quibus gignatur.Inutile enim fere fuerit,vir-
tutem quidem noffe, acquirendæ autem eius modos & vias
ignorare Non enim de virtute tantum,qua fpecie fit , quæ-
rendum eft ,fed & quomodo fui copiam faciat, vtrunque e-*

*vim volumus , et rem ipsam noſſe & eius compotes fieri :Hoc autem ex voto non ſuccedet, niſi ſciamus & ex quibus & quo modo .*In ſuch full wordes and with ſuch iteration doth he inculcate this part : So ſaith *Cicero* in great Commendation of Cato the ſecond,that he he had applyed him ſelf to Philoſophy. *Non ita diſputandi Cauſa,ſed ita viuendi.* And although the neglect of our tymes wherein few men doe houlde any Conſultations touching the reformation of theire life (as *Seneca* excellently ſaith, De *partibus vitæ quiſque deliberat,de ſumma nemo*) may make this part ſeem ſuperfluous:yet I muſt Conclude with that *Aphoriſm of Hypocrates,Qui graui morbo correpti dolores non ſentiunt,ijs mens ægrotat.* They neede medicine not onely to aſſwage the diſeaſe but to awake the ſenſe And if it be ſaide, that the cure of mens Mindes belongeth to ſacred diuinity,it is moſt true:But yet Morall Philoſophy may be preferred vnto her as a wiſe ſeruaunt,and humble handmaide.For as the Pſalme ſaith, *That the eyes of the handmayde looke perpetually towardes the miſtreſſe,* and yet no doubt many things are left to the diſcretion of the handmayde,to diſcerne of the miſtreſſe will. So ought Morall Phiꞇoſophy to giue a conſtant attention to the doctrines of Diuinity, and yet ſo as it may yeeld of her ſelfe(within due limits)Many ſoůd and profitable directions.

This *Part* therefore,becauſe of the excellency therof,I cannot but find exceeding ſtrange , that it is not reduced to written enquiry, the rather becauſe it conſiſteth of much matter,wherein both ſpeech and action
on

on is often conuerfant,and fuch wherein the common talke of men (which is rare, but yet commeth fome-times to paffe)is wifer then their Bookes : It is reafonable therefore that we propound it in the more particularity,both for the woorthineffe, and becaufe we may acquite our felues for reporting it deficiēt,which feemeth almoft incredible,and is otherwife conceiued and prefuppofed by thofe themfelues, that haue written. We wil therfore enumerate fome heads or Points thereof, that it may appeare the better what it is, and whether it be extant.

Firft therefore in this,as in all things,which ar practicall,we ought to caft vp our account,what is in our power,and what not:for the one may be dealte with by waye of alteration, but the other by waye of application onely.The hufbandman cannot command, neither the Nature of the Earth , nor the feafons of the weather: no more can the Phyfition the conftitution of the patiente,nor the varietye of Accidentes. So in the Culture and Cure of the mynde of Man, two thinges are without our commaund:Poyntes of Nature, and pointes of Fortune. For to the *Bafis* of the one,and the Conditions of the other, our worke is limited and tied. In thefe thinges therefore, it is left vnto vs , to proceede by application , *Vincenda eft omnis fortuna ferendo* : and fo likewife *vincenda eft omnis Natura ferendo* . But , when that wee fpeake of fufferinge , wee doe not fpeake of a dull , and neglected fufferinge , but of a wife and induftrious fufferinge , which draweth,

and

and contriueth vſe and aduantage out of that which
ſeemeth aduerſe and contrary; which is that property
which we cal, Accomodating or Applying. Now the
wiſedome of Application reſteth principally in the
exact & diſtinct knowledge of the precedent ſtate, or
diſpoſition, vnto which we do apply : for we cannot
fit a garment, except wee firſt take meaſure of the
Body.

So then the firſt Article of this knowlede is to
ſet downe Sound and true diſtributions and deſcriptions of the ſeueral characters & tempers of mens Natures and diſpoſitions ſpecially hauing regard to
thoſe differences which are moſt radicall in being the
fountayns and Cauſes of the reſt or moſt frequent in
Concurrence or Commixture; wherein it is not the
handling of a fewe of them in paſſage the better to
deſcribe the Mediocrities of vertues that can ſatiſfie
this intention for if it deſerue to be conſidered *That
there are minds which are proportioned to great mattes, &
others to ſmal,* (VVhich Ariſtotle handleth or ought to
haue handled by the name of Magnaminity) doth it
not deſerue as well to be Conſidered. *That there are
mindes proportioned to intend many matters and others to
few?* So that ſome can deuide them ſelues others can
perchance do exactly wel, but it muſt bee but in fewe
things at once; And ſo there cometh to bee a *Nar-
rownes of mind* as wel as a *Puſillanimity.* And againe,
*That ſome mindes are proportioned to that which may bee
diſpatched at once or within a ſhort return of time: others to
that which begins a farre of, and is to be won with length of*
pur.

purſute ,—Iam tŭ tenditque ſouetqne; So that there may
be fitly ſaid to be a longanimity which is Comonly
alſo aſcribed to God as a *Magnanimity* So further
deſerued it to be conſideted, by Ariſtotle *That there
is a diſpoſition in Conuerſation* (*ſuppoſing it in things
which doe in no ſort tonch or concerne a mans ſelfe*) *to ſoothe
aud pleaſe*; *And a diſpoſition contrary to Contradict and
Croſſe*; And deſerueth it not much better to be conſidered,*That there is a diſpoſition, not in conuerſation or
talke, but in matter of mere ſerious Nature (and ſuppoſing
it ſtill in things mœerly indifferent*) *to take pleaſure in the
good of another, and a diſpoſition contrarywiſe, to take diſtaſt
at the good of another*;which is that properly,which we
call good Nature,or ill Nature, benignity or Malig-
nity:And therefore I cannot ſufficiently Maruayle ,
that this parte of knowledge touching the ſeuerall
Characters of Natures and diſpoſitions ſhould bee
omitted both in Morality and policy,conſidering it is
of ſo great Miniſtery,and ſuppeditation to them both
A man ſhall find in the traditions of Aſtrology,ſom
prety and apt diuiſions of mens Natures according
to the predominances of the Planets; *Louers of Quiet
Louers of action, louers of victory, loucrs of Honour, louers
of pleaſure, loucrs of Arts, louers of Change,*and ſo forth:
A man ſhall find in the wiſeſt ſort of theſe Relations,
which the Italians make touching *Conclaues,*the Natures of the ſeuerall Cardinalls , handſomlye and
liuely painted fourth : A man ſhall meete with in e-
uery dayes Conference the denominations *of Serſi-
tiue, dry, formall, reall, humorous, certayne, Humo di Prima
impreſſione, Huomo di vltima impreſſione,* and the like,
and

and yet neuerthelesse this kind of obseruations wan-dreth in wordes,but as not fixed in Enquiry.For the distinctions are found (many of them) but we con-clude no precepts vpon them, wherein our faulte is the greater,becaufe both History, Poefye,and daylie experience are as goodly fields where thefe obfer-uations grow, whereof wee make a few poefies, to hould in our hands,but no man bringeth them to the confectionary,that Receits mought be made of them for vfe of life.

Of much like kinde are thofe impreffions of Nature,which are impofed vpon the Mind *by the Sex, by the Age, by the Region, by health, and ficknesse* , by *beauty* and *deformity*, and the like , which are inhe-rent,and not externe:and again thofe which are cau-fed by extern fortune:as *Soueraygnty, Nobility, obfcure birth,ritches, want,Magiftracye, priuatenesse, profperity, aduerfity, Conftant fortune, variable fortune, rifing per faltum,per gradus* , and the like : And therefore we fee,that *Plautus* maketh it a wonder,to fee an oulde man beneficent, *Benignitas huius vt adolefcentuli eft :* Saint *Paul* concludeth that feuerity of difcipline was to be vfed to the Cretans, *Increpa eos durè,* vpon the difpofition of their Country *Cretenfes femper mendaces, malæ Beftiæ, ventres pigri* . *Saluft* noteth, that it is vfuall with Kinges to defire Contradic-toryes,*fed plerunque Regiæ voluntates, vt vehementes funt, fic mobiles,fæpeque ipfæ fibi aduerfæ* . *Tacitus* ob-
                                                                                      ferueth

serueth how rarely raising of the fortune mendeth the disposition , *solus Vespasianus ;mutatus in melius ,* *Pindarus* maketh an obseruation , that greate and suddaine fortune for the most parte defeateth men *Qui magnam fœlicitatem concoquere non possunt* : So the Psalme sheweth it is more easie to keep a mea˜sure in the enioying of fortune , then in the increase of fortune . *Diuitiæ si affluant, nolite Cor ap˙ponere :* These obseruations and the like , I denye not, but are touched a little by *Aristotle* as in passage in his Rhetoricks , and are handled in some scattered discourses , but they were neuer incorporate into Morall Philosophy, to which they doe essentiallye appertayne: as the knowledg of the diuersitye of groundes and Mouldes doth to *Agri*culture, and the knowledge of the diuersity of Complexions and Constitutions doth to the Phisition; except we meane to follow the indiscretion of Empe riques,which minister the same medicines to all patients.

Another Article of this knowledge is theInquirye touching the affections : for as in Medicining of the body it is in order first to know the diuers Complexions and constitutions , secondlye the diseases , and lastlye the Cures : So in medicining of the Minde , after knowledge of the diuers Charact-ers of mens natures ,it foloweth in order to know the diseases and infirmites of the mind, which ar no o=ther the˜ the perturbations & distemper s of the affec-

tions

tions. For as the aunciente in politiques in po͞-
pu!er Eſtates were woont to to Compare the people to the ſea, and the Orators to the winds becauſe
as the ſea would of it ſelfe be caulm and quiet, if the
windes did not mooue and trouble it; ſo the people
would be peaceable and tractable if the ſeditious ora
tors did not ſet them in working and agitation. So it
may be fitly ſaid, that the mind in the nature thereof
would be temperate and ſtayed, if the affections as
winds, did not put it into tumulte and perturbation.
And here againe I find ſtraunge, as before, that *A-
riſtotle* ſhoulde haue written diuers volumes of E-
thiques, and neuer handled the affections, which is
the principall ſubiect thereof, and yet in his Retoricks
where they are conſidered but collaterally, & in a ſe-
cōd degree, *(as they may be mooued by ſpeech)*he findeth
place for them, and hādleth them well for the quātity
but where their true place is, he pretermitteth them.
For it is not his diſputations about pleaſure and paine
that can ſatiſſie this inquiry, no more then hee that
ſhould generally handle the nature of light can bee
ſaid to handle the nature of *Colours*: for pleaſure and
paine are to the particuler affections as light is to
particular collours : Better trauailes I ſuppoſe had
the *Stoicks* taken in this argument, as far as I can gather by that which wee haue at ſecond hand: But yet
it is like, it was after their manner rather in ſubtiltye
of definitions (which in a ſubiect of this nature are
but curioſities)then in actiue and ample deſcriptions
and obſeruations: ſo likewiſe I finde ſome particular
wri-

writings of an elegant nature touching some of the affections , as of *Anger*, of *Comforte vpon aduerse accidentes* , of *Tendernesse of Countenance* and other. But the poets and wtiters of Histories are the best Doctors of this knowledge, where we may finde painted fourth with greate life, How affections are kindled and incyted:and how pacified and restrai ned: and how againe Conteyned from *Act*, & furder degree:how they disclose themselues,how they work how they varye,how they gather and fortifie, how they are inwrapped one within another, aud howe they doe fighte and encounter one with another , and other the like particularityes: Amōgst the which this last is of speciall vse in Morall and Ciuile matters: I owe I say to sett affection againste affection, and to Master one by another,even as wee vse to hunt beast with beaste, aud flye byrde with birde, which otherwise percase wee coulde not so easily recover: vpon which foundation is erected that excellent vse of *Præmium* and *pœna*. whereby Ciuile states Consist, imploying the predominante affections of *feare* and *hope*, for the suppressing and brideling the rest. For as in the gouernemente of states, it is sometimes necessarye to bridle one faction with another, so it is in the gouernmente within.

Now Come we to those poynts which are within our our owne cōmand and haue force and operacion vpon the mind to affect the wil & Appetite & to alter Manners:wherin they ought to haue hādled *Custome*

Exercife, *Habit*, *Educacion*, *example*, *Imitation*, *Emulation*
Company, *Frinds*, *praife*, *Reproofe*, *exhortatiō*, *fame*, *lawes*,
Bookes, *ftudyes*: theis as they haue determinate vfe,
in moralityes, from thefe the mind fuffereth, and of
thefe are fuch receipts & Regiments compounded &
defcribed, as may feeme to recouer or preferue the
health and Good eftate of the mind, as farre as pertaineth to humane Medycine: of which number wee
will vifit vpon fom one or two as an exāple of the reft,
becaufe it were too ong to profecute all; and therefore wee doe refume Cuftome and habite to fpeake
of.

   The opinion of Arifto:le feemeth to mee a negligent opinion. That of thofe thinges which confift by
nature, nothing can be changed by cuftome, vfing for
example: That if a ftone bee throwne ten thoufande
tymes vp, it wil not learne to affend, and and that by,
often feeing or hearing, wee doe not learne to fee or
heare the better. For thoughe this principle bee true
in things wherein nature is *Peremptory* (the reafon
whereof we cannot nowe ftande to difcuffe) yet it is
otherwife in things wherein nature admitteth *a latitude*. For he moughtfee that a ftreight gloue wil come
more eafily on with vfe, and that a wand will by vfe
bend otherwife then it grewe: and that by vfe of the
voice wee fpeake lowder and ftronger , and that
by vfe of enduring heate or coulde, we endure it the
better, and the like. which later fort haue a neerer refemblance vnto that fubiect of Manners he handleth
then thofe inftāces which he alledgeth; But allowing
                                                  his

his Conclusion *that vertues aud vices confift in habit,* he ought fo much the more to haue taught the manner of fuperinducing that habite:for there bee many precepts of the wife ordering the exercifes of the minde , as there is of ordering the exercifes of the body, wherof we wil recite a fewe.

The firft fhal bee, that wee beware wee take not at the firft either to *High* a ftrayne or to *weake*:for if, too *Highe* in a differēt nature you difcorage, in a confident nature,you breede an opinion of facility,and fo a fiorh, and in all natures you breede a furder expectation then can hould out,and fo an infatisfaction on the end,if to weake of the ether fide :you may not looke to performe and ouercome any great talke.

Another precept is to practife all thinges chiefly at two feuerall times, the one when the mind is befte difpofed,the other when it is *worfte difpofed*:that by the one you may gaine a great ftep,by the other you may worke out the knots and Stondes of the mind, and make the middle times rhe more eafily and pleafant.

Another precept is,that which Ariftotle mencio, neth by the way,which is to beare euer towards the Contrary extreame of that wh erevnto we are by Nature inclyned:like vnto the Rowing againft the ftream or making a wand ftraight by bynding him Contrary to his natural Crookedneffe.

Another precept is,that the mind is brought to any thing better and with more fweetneffe and happineffe,if that wherevnto you pretend be not firft in the
intention

intention. but *Tanquā aliud agendo*, becaufe of the Na
turall hatred of the minde againft neceffity and Con
ftraint. Many other Axiomes there are touching the
Managing of *Exercife* and *cuftome*: which being fo
Conducted, doth prooue indeed another nature: but
being gouerned by chance, doth cōmōly prooue but
an ape of nature, & bringeth forth that which is lame
and Counterfette.

So if wee fhoulde handle *bookes* and *ftudies* and
what influence and operation they haue vpon manners, are there not diuers precepts of greate caution
and direction appertaining thereunto? did not one
of the fathers in greate indignation call *Poefy vinum
Demonum*, becaufe it increafeth temptations, perturbations and vaine opinions? Is not the opinion of *A-
riftotle* worthy to be regarded wherein he faith, That
yoūg men are no fitte auditors of Moral Philofophy,
becaufe they are not fetled from the boyling heate
of their affections, nor attempered with *Time* and
experience? and doth it not hereof come that thofe
excellent books and difcourfes of the aunciente
writers, (whereby they haue perfwaded vnto *vertue*
moft effectually, by reprefenting her in *ftate* and *Ma=
iefty*, and *populer opinions* againft vertue in their *Parafites Coates*, fitt to be fcorned and derided,) are of fo
little effect towards honefty of life, becaufe they are
not red & reuolued by mē in their mature and fetled
yeares, but confined almoft to boyes & beginners? But
is it not true alfo that much leffe, young men are
fit auditors of Matters of Policy, till they haue beene
throughly

throughly seasoned in religion & Morality, least their
Iudgementes be corrupted, and made apt to thinke
that there are no true Differences of things, but according, to *vtility* and *fortune*, as the verse describes it.
Prosperum et Fœlix scelus virtus vocatur: And Againe
Ille crucem pretium sceleris tulit, Hic diadema: which the
Poets do speak satyrically and in indignation on vertues behalfe: But books of pollicie doe speake it seri-
oufly, and positiuely, for so it pleaseth *Machiauell* to
say *That if Cæsar had bene ouerthrowne, he woulde haue
beene more odious then euer was Catiline*; as if there had
beene noe difference but in fortune, between *a very
fury of luft & bloud*, and *the moft excellët spirit* (his am-
bicio referued) *of the werld*? Again is there not a Caution likewise to be giuen of the doctrines of Moralities thefelues (some kindes of thë) leafte they make
men too precife, arrogãt, incõpatible, as *Cicero* faith
of *Cato in Marco Catone. Hæc bona quæ videmus diuina &
egregia ipfius scitote effe propria: quæ nonnunquam requi-
rimus, eas funt omnia, non a natura fed a Magiftro?* Many
other Axiomes & aduifes there are touching thofe
proprieties & effects, which ftudies doe infufe & inftil
into maners: And fo likewife is there touchinge the
vfe of all thofe other points of Company fame, lawes
and the reft, which we recited in the beginning in
the doctrine of Morality.
But there is a kind of C v l t v r e of the M i n d; that
femeth yet more accurate & elaborate thë the reft &
is built vpon this ground: That the minds of all men
are at fome times in a ftate more perfite, and at o-
ther

*other tymes in a state more depraued.*The purpose therfore of this practise is to fixe and cherishe the good howers of the mind and to obliterate and take fourth the Euil:The fixing of the good hath bene practised by two meanes,vowes or Constant resolutions, and obseruances,or exercifes which are not to be regar. ded so much in themselues,as becaufe they keepe the mynd in continual obedience.The obliteratiō of the Euill hath been practised by two Meanes,some kind of Redemption or expiation of that which is past, and an Inception or account *de Nouo,* for the time to come: but this part, seemeth sacred and religious, and Iustly: for all good Morall Philofophy(as was said,)is but an handmaide to Religion.

Wherefore we will conclude with that laft pointe which is of all other meanes the moste compen. dious aud summarye, and againe, the moste noble and effectual to the reducing of the minde vnto ver. tue and good eftate:which is the electing and pro. pounding vnto a mans felfe good & vertuous ends of his life,such as may bee in a reasonable forte within his Compas to attaine.For if these two things be sup osed: that a mã fet before him honeft & Good ends, and againe that he bee refolute, Conftant, and true vnto them; it will follow that hee shall Moulde him. selfe into al vertue at once: and this is indeede like the worke of nature, whereas the other courfe , is like the worke of the hand. For as when a caruer makes an image ,hee shapes onely that parte whereupon hee worketh,as if hee bee vpon *The face* that parte which shall

ſhal bee *the body* is but a rude ſtone ſtil, til ſuch times
as hee comes to it . But contrarywiſe when Nature
makes a *flower* or *liuing creature*, ſhee ſourmeth rudiments of all the parts at one time; ſo in obtaining ver
tue by *habite*, while a man practiſeth Temperance, he
doth not profit much to fortitude, nor the like; But
when he dedicateth & applyeth himſelfe to *good ends*,
loke what vertue ſoeuer the purſute and paſſage towards thoſe ends doth commend vnto him , he is in-
ueſted of a precedent diſpoſition to conforme himſelfe thereunto: which ſtate of mind *Ariſtotle* doth ex.
excellently expreſſe himſelf, that it ought not to bee
called *vertuous*, but *Diuine*: his words are theſe; *Imma.
nitati autem conſentaneum eſt, opponere eam, quæ ſupra
humanitatem eſt, heroicā ſiue diuinam virtutem.* And a
little after; *Nã vt fera, neque vitiū, neq; virtus eſt ſic neq;
Dei. Sed hic quiãe ſtatus altius quiddã virtute eſt, itie aliud
quiddã a vitio.* And therfore we may ſee what Celſitud
of honor *P.inius ſecundus* attributeth to *Traiane* in his
funerall oration, where he ſaid . *That men needed to
make nce other praiers to the Gods, but that they woulde
Continue as good Lords to them, as Traiane had beene:* as
if he had not beene onely an Imitation of diuine nature , but a patterne of it. But theſe be heathen & prophane paſſages having but a ſhadowe of that diuine
ſtate of mind, which Religion and the holy faith doth
conduct men vnto ; by imprinting vpon their ſoules
Charity which is exellctly called the bond of *Perfectiõ*:
bicauſe it cõprehẽdeth & faſtneth al vertues together.
And it is elegantly ſaid by *Menander* of vaine loue
which

which is but a falſe Imitation of diuine *loue. Amor melior Sophiſta, Lauo ad humanam vitam*, that Loue teacheth a man to Carry himſelfe better, then the *Sophiſt* or *Præceptor*, which he calleth *Leſt handed*, becauſe with all his rules & preceptiõs he cannot form a man ſo *Dexteriouſly*, nor with that facility to prize himſelf & gouern himſelf as loue cã do: So certainly if a mãs mind be truly inflamd with charityit doth work himſodainly into greter perfectiõ then al the Doctrin of moralitye can doe, which is but a ſophiſt in compariſon of the other. Nay furder as *Xenophon* obſerved truely that all other affections though they raiſe the minde, yet they doe it by diſtorting, and vncomlineſſe of extaſies or exceſſes; but onely Loue doth exalt the mind, and neuertheleſſe; at the ſame inſtant doth ſettle and *Compoſe* it. So in all other excellencyes though they aduance nature yet they are ſubiect to Exceſſe. Onely Charity admitteth noe *Exceſſe*; for ſoe we ſee, aſpiring to be like God in power, the Angells tranſgreſſed and fel: A*ſcendam, & ero ſimilis altiſſimo*: By aſpiringe to be lſke God in knowledge man tranſgreſſed and fell. *Eritis ſicut Dii ſcientes bouum & malum*; But by aſpiring to a ſimilitude of God in goodneſſe or loue, neyther Man nor Angell euer tranſgreſſed or ſhall tranſgreſſe. For vnto that imitation wee are called, *Diligite inimicos veſtros, Beneſacite eis qui odernnt vos, & orate pro perſequentibus & Calumniantibus vos vt ſitis filii patris veſtri qui in cælis eſt, qui ſolem ſuum oriri facit ſuper bonos*

bonos & malos, & pluit super iustos & iniustos. So in the first platfourme of the diuine Nature it self, the heathē Religion speaketh thus, *Optimus Maximus,* and the sacred scriptures thus, *Misericordia eius super omnia opera eius.*

Wherefore I doe conclude this part of Morall knowledge concerning *the Culture and Regiment of the Mind,* wherin if any man considering the parts therof, which I haue enumerated, doe iudge, that my labor is but to Collect into an *Art* or *Sciēce,* that which hath bin pretermitted by others, as matter of cōmon *Sence, and experience,* he iudgeth well: But as *Philocrates* sported with *Demosthenes : you may not maruaile (Atheuians)* that *Demosthenes and I doe differ, for h:e drink-e.h water, and I drinke wine* : and like as wee reade of an aunciente parable of *the two gates of sleep.*

Sunt geminæ somni portæ, quarum altera fertur
Cornea, qua veris facilis datur exitus vmbris:
Altera Candenti perfecta nitens Elephanto,
Sed falsa ad cælum mittunt insomnia manes.

So if wee put on sobriety and attention, we shall finde it a sure Maxime in knowledge : that the more pleasaunte Liquor *(of Wine)* is the more vaporous, and the brauer gate *of Iuorye,* sendeth foorthe the falser dreames.

But we haue now concluded, *That generall part of Humane Philosophye, which contemplateth man segregate, and as hee consisteth of bodye and spirite ;* Wherein wee maye further note, that there

there feemeth to be a Relatiō or Conformity betwen the good of the mynd, and the good of the Body. For as we devided the good of the body into *Health, Beau ty, ſtrength,* and *Pleaſure,* fo the good of the mynde inquired in Rationall and Morall knoweledges tendeth to this, to make the minde found, and without perturbation, *Beautifull* and graced with decencie: and *Strong* and *Agill* for all duties of life. Theis three as in the bodye, fo in the minde feeldome meete, and Commonly feuer : For it is eafilye to obferue, that many haue Strength of witte and Courage, but haue neither Healthe from perturbations, nor any Beauty or decencie in theire doings: fom againe haue an Elegancy and fineneſſe of Carriage, which haue neither foundneſſe of honeſtie, nor fubſtance of fufficiencye : And fome againe haue honeſt and refourmed *My*ndes that can neither become themfelues nor Manage Bufineſſe, and fometimes two of them meete, and rarely all three: As for pleafure, wee haue likewife determined, that the minde oughte not to bee reduced to ftupide, but to retayne pleafure: Confined rather in the fubieĉt of it, then in the ftrength and vigor of it.

Civile Knovvledge is converfant about a fubiect which of all others is moſt immerfed in matter, and hardlieſt reduced to Aĉtiome. Neuertheleſſe

Neuerthelesse, as *Cato* the Censor saide, *That the Romane: were like sheepe , for that a man were better driue a flock of them ,then one of them; For in a flocke, if you could get but some fewe goe righte,the rest would follow:* So in that respect Morall philosophie is more difficile then Pollicie. Againe , morrall Philosophye propoundeth to it selfe the framing of Internall goodnesse: But ciuile knowledge requireth onelye an Externall goodnesse : for that as to societye sufficeth:And therfore it cometh oft to passe that therebe Euill Times in good gouernments:for so we finde in the holy story when the kings were good,yet it is ad. ded.*Sed adhuc populus non dixerat cor suum ad dominum Deum patrum suorum.*Againe States as great Engines mooue slowly,and are not so soone put out of frame:for as in *Ægypt* the seauē good years sustained the seauen badde: So gouernments for a time well grounded doe beare out errors following. But the resolution of particuler persons is more sodainly subuerted,These respects doe somwhat qualfie the extreame difficulty of ciuile knowledge.

This knowledge hath three parts according to the three summary Actiōs of society, which are,Cōuersation,Negotiatiō aud Gouernment. For mā seeketh in society comfort,vse and Protection:& theybe three wisedōs of diuers natures,which do oftē seuer: wisedome of the behauiour, wisedom of Businesse;& wisedome of *state.*

The wisedome of of conuersation ought not to be ouer mvch affected, bnt much lesse despised: for it

hath

hath not onely an honour in it felfe, but an influence alfo into bufineffe and gouernment; The poet faieth.

Nec vultu deſtrue verba tuo · A man maie deſtroy the force of his woords with his countenance: fo may he of his deeds faieth *Cicero,* recommending to his brother affability and eaſy acceſſe, *Nil intereſt habere oſtium apertum, vultum clauſum.*

It is nothing wonne to admitte men with an open doore, and to receiue them with a ſhutte and reſerued countenaunce · So wee fee *Atticus,* before the firſt interuiewe betweene *Cæſar* and *Cicero,* the warre depending, did ſeriouſlye aduiſe *Cicero* touching the compoſing and orderinge of his countenaunce and geſture. And if the gouernemente of the countenaunce bee of ſuch effecte, much more is that of the ſpeeche, and other carriage appertayning to conuerſation; the true modele whereof ſeemeth to mee well expreſſed by *Liuye,* though not meante for this purpoſe; *Ne aut arrogans videar, aut obnoxius, quorum alterum eſt alienæ libertatis obliti, alterum ſuæ:* The ſumme of behauioure is to retayne a mans owne dignitye, without intruding vpon the libertye of others : on the other ſide, if behauioure and outwarde carriage bee intended too much, firſt it may paſſe into affection, and then *Quid deformius quam Scænam in vitam transferre,* to acte a mans life ? But although it proceede not to that extreame, yet it conſumeth time, and imployeth

eth the minde too much. And therefore as wee vſe to aduiſe younge ſtudentes from company kee‑ ping, by ſaying, *Amici, ſures Temporis* : So certainely the Intending of the diſcretion of behauioure is a great Theefe of Meditation. Againe, ſuch as are ac‑ compliſhed in that howr of vrbanity, pleaſe them‑ ſelues in name, and ſildome aſpire to higher vertue: whereas thoſe that haue defect in it, do ſeeke *Com‑ lines* by Reputation: for where reputacion is, almoſt euery thing becommeth : But where that is not, it muſt be ſupplied by *Puntos* and Complementes : A‑ gayne, there is no greater impediment of Action, then an ouercurious obſeruaunce of decency, and the guide of decencye, which is Tyme and ſeaſon. For as *Salomon* ſayeth, *Qui reſpicit ad ventos, non ſeminat, & qui reſpicit ad nubes, non metet*: A man muſt make his opportunity, as ofte as finde it. To conclude; Behauiour ſeemeth to me as a Garment of the Minde, and to haue the Condicions of a Garmente . For it ought to bee made in faſhi‑ on : it ought not to bee too curious : It ought to bee ſhaped ſo , as to ſette foorthe anye good making of the minde: and hide any deformity; and aboue all, it ought not to be too ſtraighte, or re‑ ſtrayned for exerciſe or mocion . But this parte of Ciuile knowledge hath beene elegantlye hand‑ led, and therefore I cannot reporte it for defici‑ ent.

The wiſedome touching Negotiation or buſineſſe

hath

hath not bin hitherto collected into writing to the great derogacion of learning, and the profeſſors of learninge. For from this roote ſpringeth chiefly that note or opinion which by vs is expreſſed in A. dage, to this effecte: That there is noe greate con. currence betweene learning and Wiſedome. For of the three wiſedomes which wee haue ſette downe to pertaine to ciuil life, for wiſedome of Behauiour, it is by learned men for the moſte parte deſpiſed, as an Inferiour to Vertue and and an Enemy to Medi-tacion; for wiſedome of Gouernmente they acquite themſelues well when they are called to it, but that happeneth to fewe. But for the wiſedome of Buſi-neſſe wherein mans life is moſte conuerſant, there bee noe Bookes of it, excepte ſome fewe ſcattered aduertiſementes , that haue noe proportion to the magnitude of this ſubiecte. For if bookes were writ. ten of this, as the other, I doubt not but learned men with meane experience, woulde farre excell men of longe experience withoute learning, and out-ſhoote them in their owne bowe.

Neither needeth it at all to be doubted, that this knowlddge ſhoulde bee ſo variable as it falleth not vnder precept; for it is much leſſe infinite then ſcience of Gouernmente, which wee ſee is laboured and in ſome parte reduced. Of this wiſedome it ſee-meth ſome of the auncient Romanes in the ſaddeſt and wiſeſt times were profeſſors : for Cicero repor-teth, that it was then in vſe. For Senators that had

canius

name and opinion, for generallwife men as *Corun-caniuus*, *Curius*, *Lælius* and manie others; to walke at certaine howers in the *Place*, and to giue audience to thofe that would vfe their aduife, and that the particuler Citizens would refort vnto them, and con, fulte with them of the marriage of a daughter, or of the imploying of a fonne, or of a purchafe or bar-gaine, or of an accufatiõ and euery other occafion incident to mans life; fo as there is a wifedome of Counfaile and aduife euen in priuate Caufes: arifinge out of an vniuerfall infight into the affayrs of the world, which is vfed indeede vpon particuler cafes propoũded but is gathered by generall obfer, uation of caufes of like nature. For fo wee fee in the Booke which Cicero writeth to his brother *De petitione confulatus*, (being the onely booke of bufineffe that I know written by the auncients) although it cõcerned a particuler action then on foote, yet the fub-ftance thereof confifteth of manie wife and polli-tique Axioms which containe not a temporary, but a perpetuall direction in the cafe of popular Electi-ons; Eut chiefly wee may fee in thofe Aphorifmes which haue place amongeft Divine writings com-pofed by Salomon the King, of whom the fcriptures teftifie that his hearte was as the fandes of the fea, in-compaffing the world and all worldly matters we feeI faie, not a few profound and excellent cauti-ons, precepts, pofitions, extending to mnch varie-tie of occafions; wherevpon wee will ftaie a while

offering

offering to confideracion fome number of Examples.

*Sed & cunctis fermonibus qui dicuntur, ne accom-
modes aurem tuam, nè fortè audias feruum tuum male-
dicentem tibi .* Heere is concluded the prouidente
ftaye of enquiry, of that which we wolde be loathe
to finde : as it was iudged greate wifedome in *Pompeius Magnus* that he burned *Sertorius* papers vnper-
ufed.

*Vir fapiens fi cum ftulto contenderit, fiue irafcatur, fiue
rideat, non inueniet requiem.* Here is defcribed the great
difaduantage which a wife man hath in vndertaking
a lighter perfon then himfelfe, which is fuch an in-
gagemente, as whether a man turne the matter to
ieaft, or turne it to heate; or howfoeuer hee change
copye, hee can no wayes quitte himfelfe well of
it.

*Qui delicatè à pueritia nutrit feruum fuum, poftea
fentiet eum contumacem .* Heere is fignified that if a
man beginne too highe a pitche in his fauoures, it
doeth commonlye end in vnkindneffe, and vnthank-
fulneffe.

*Vidifti virum velocem in opere fuo, coram regibus ftabit
nec erit inter ignobiles.* Here is obferued that of all vertues for rifing to honoure, quickneffe of difpatche is
the beft; for fuperiours many times loue not to haue
thofe they imploy too deep, or too fufficient, but redy
and diligent.

*Vidi cunctos viuentes, qui ambulant fub fole cum ado-
lefcente*

adolescente secundo qui consurgit pro eo. Here is expreffed that which was noted by Sylla firft, and after him by *Tiberius; Plures adorant solem orientem, quam occidentem vel meridianum.*

Si spiritus potestastem habentis ascenderit super te, locum tuum ne dimiseris, quia Curatio faciet cessare peccata maxima. Here caution is giuen that vpon displeafure, retiring is of all courfes the vnfitteft; for a man leaueth thinges at worft, and depriueth himfelfe of meanes to make them better.

Erat Ciuitas parua pauci in ea viri; venit contra eam rex magnus, & vadauit eam, insirxuitque munitiones per Gyrum, & perfecta est obsidio inuentusquæ est in ea vir pauper & sapiens, & liberauit eam per sapientiam suam, & nullus aeinceps recordatus est hominis illius pauperis; Here the corruptions of ftates is fette foorh; that efteeme not vertue or merite longer then they haue vfe of it.

Mollis responsio frangit iram. Here is noted that filence or rough Anfweare, exafperateth: but an anfwear prefent and temperate pacifieth.

Iter pigrorum, quasi sepes spinarum. Here is liuelie reprefented how laborious floth prooueth in the end; for when thinges are differred till the lafte inftant, and nothing prepared before hande, euerye ftepp findeth a Bryer or Impediment, which catcheth or ftoppeth.

Melior est finis orationis quam principium. Here is taxed the vanitie of formall fpeakers, that ftudy more

about

about prefaces and inducements, then vpon the conclufions and iffues of fpeache.

Qui cognofcit in iudicio faciem, non bene facit, iste et pro buccella panis deferet veritatem. Here is noted that a iudge were better be a briber, then a refpecter of perfons :for a corrupt Iudge offendeth not fo lightly as a facile.

Vir pauper calumnians pauperes, fimilis eft imbri vehementi, in quo paratur fames; here is expreffed the extreamiiy of neceffitous extortions, figured in the auncicnte fable of the full and the hungry horfeleech.

Fons turbatus pede, & vena corrupta, eft iuftus cadens coram impio: here is noted that one iudiciall and exemplar iniquity in the face of the world, doth trouble the fountaines of Iuftice more, then many particuler Iniuries paffed over by conniuence.

Qui fubtrahit aliquid a patre & a matre, & dicit hoc non effe peccatū, particeps eft homicidij; here is noted that whereas men in wronging theyr beft frindes, vfe to extenuat: their faulte, as if they moughte prefume or bee bo'de vpon them, it doth contrariwife indeede aggrauate their fault, & turneth it from Iniury to impiety.

Noli effe amicus homini iracundo, nec ambulato cum homine furiofo; here caution is giuen that in the election of our friends wee doe principalliy avoide thofe which are impatiente, as thofe that will efpoufe vs to many factions and quarels.

Qui conturbat domum suam possidebit ventum:
here is noted that in domesticall separations & brea-
ches men doe promise to themselues quietting of
theire minde and contentemente, but still they are
deceiued of theire expectation, and it turneth to
winde.

Filius sapiens lætificat patrem, filius vero stultus mæsti-
cia est matri suæ. Here is distinguished that fathers
haue moste comforte of the good proofe of
of their sonnes ; but mothers haue moste discomfort
of their ill proofe, because women haue little discer-
ninge of vertue but of fortune.

Qui celat delictum quærit amicitiam, sed qui altero ser-
mone repetit, seperat fæderatos ; here caution is giuen
that reconcilemente is better managed by an *Am-*
nesty and passing ouer that which is past, then by A-
pologies and excusations.

In omni opere bono erit abundantia, vbi autem verba
sunt plurima, ibi frequenter egestas : here is noted that
words and discourse aboundeth moste, where there is
idlenesse and want.

Primus in sua causa iustus, sed venit altera pars, & in-
quiret in eum; Heare is obserued that in all causes the
first tale possesseth much, in sorte, that the preiudice,
thereby wrought wil bee hardly remooued, excepte
some abuse or falsitie in the Information be detected.

Verba blinguis quasi simplicia, & ipsa peruentunt ad
interiora ventris ; there is distinguished that flatterye
and insinuation which seemeth set and artificiall,

A a a 2   sinketh

finketh not farre, but that entreth deepe, which hath
fhewe of nature, libertie, and fimplicity,

Qui erudit derifo: em ,ipfe t bi inuria n facit &
qui arguit Impium fibi maculam gerit. Here caution
is giuen howe wee tender reprehenfion to arro-
gante and fcornefull natures, whofe manner is to
efteeme it for contumely, and accordingly to retourne it,

Da fapienti occafionem & addetur ei fapientia,
Here is diftinguifhed the wifedome, broughte into
habite, and that which is but verball and fwimming
onely in conceite: for the one vpon the occafione
prefented is quickned and redoubled: the other is a-
mazed and confufed.

Quo modo in aquis refplendent vultus profpicien-
tium, fic corda hominum manifefta sunt prudentibus.
Here the mind of a wife man is compared to a glaffe,
wherein the Images of all diuerfitie of Natures
& Cuftoms are reprefeted, frō which reprefentatiō
proceedeth that application,
Qui sapit innumeris moribus aptus erit,

Thus haue I ftaide fome what longer vpon
thefe fentences pollitique of Salomon, then is agre,
able to the proportion of an example: ledde with
a defire to giue authority to this parte of knoweledge, which I noted as deficiente by fo excel-
lente a prefidente: and haue alfo attended them
with briefe obferuations, fuch as to my vnder-
ftandinge, offer noe violence to the fence, though
I knowe they may bee applyed to a more diuine vfe:

but

But it is allowed euen in diuinity, that some Interpretations, yea and some writings haue more of the *Eagle*, then others: But takinge them as *Instructions* for life, they moughte haue receiued large discourse, if I woulde haue broken them and illustrated them by diducements and examples.

Neither was this in vse only with the hebrews, But it is generally to be found in the wisdome of the more auncient Times: that as men founde out any obser. uatiō that they thought was good for life, they would gather it and expresse it in parable, or *Aphorisme*, or fable. But for fables they were vicegerents & supplies, where Examples failed: Nowe that the times abounde with historie, the Ayme is better when the marke is aliue. And therefore the fourme of writing which of al others is fittest for this variable argumente of Negotiation and occasions is that which *Machiauel* chose wisely and aptly for Gouernmente: *namely discourse vpon Histories or Examples.* For knoweledge drawne freshly and in our view out of particulers, knoweth the waie best to particulers againe. And it hath much greater life for practise: when the discourse attendeth vpon the Example, then when the example attenddeth vpon the discourse. For this is no pointe of order as it seemeth at firste but of substance. For when the Example is the grounde being set downe in an history at large, it is set down with al circumstā-ces: which manye sometimes controul the discourse
thereupon

thereupon made, and fome times fupply it; as a verie patterne for gaine; wheras the Examples alledged for the difcourfes fake, are cited fuccinctly, and with out particularity, and carry a feruile afpecte towards the difcourfe, which they are broughte in to make good.

But this difference is not amiffe to bee remembred, that as hiftorye of *Tymes* is the beft grounde for difcourfe of Gouernemente, fuch as *Machyauel* handleth ; fo Hiftories of Liues is the mofte proper for difcourfe of bufineffe is more converfante in priuate Actions. Nay, there is a ground of difcourfe for this purpofe, fitter then them both which is *difcourfe vpon letters*, fuch as are wife and weightie, as manie are of *Cicero ad Atticum* and others .For letters haue a greate and more particuler reprefentation of bufineffe, then either *Chronicles* or *Liues*, Thus haue wee fpoken both of the matter and fourme of this parte of Ciuile knowledge touching Negotiation, which wee note to be deficient .

But yet there is another part of this part, which differeth as much frō that wherofwe haue fpokē as *fapere*, & *fibi Sapere:* the one moouing as it were to the circū rence, the other to the center: for there is a wifedome of counfell, and againe there is a wifedome of preffiing a mans owne fortune ; and they doe fometimes meet, and often feuere. For many are wife in their owne ways, that are weak for

gouern;

gouernmente or Counfell, like Ants which is a wife creature for it felf, but very hurtefull for the gar-den- This wifedome the Romanes did take much knoweledge of, *Nam pol fapiens* (faith the Comicall Poet) *Fingit fortunam fibi,* and it grewe to an adage, *Faber quifque fortunæ propriæ:* and *Liuie* attributeth it to *Cato* the firft , *In hoc viro tanta vis animi & ingenij inerat, vt quocunque Loco natus effet fibi ipfe fortunam facturus videre tur.*

This conceit or pofition if it bee too much declared and profeffed , hath beene thoughte a thinge impolitique and vnlucky, as was obferued in *Timotheus* the Athenian : who hauinge done manie greate feruices to the Eftate in his gouernmēt and giuinge an accounte thereof to the people as the manner was , did conclude euery particuler with this Claufe, And in this fortune had noe part And it came fo to paffe that hee neuer prospered in any thinge hee tooke in hande afterwarde: for this is too high and too arrogant fauouring of that which *Ezechiel* faith of *Pharaoh*: *Dicis: fluuius eſt meus & ego feci memet ipfum* : or of that which another prophette fpeaketh : That men offer Sacrifices to theire nettes and fnares, and that which the *Poett* expre=ffeth, *Dextra mihi Deus , & telum quod inutile libro. Nunc adfinte :*

For thefe confidences were euer vnhal-lowed , and vnbleffed. And therefore those that were great Pollitiques indeede euer afcribed their fucceffes to their felicitie : and not to theire skill or
vertue

vertue. For fo Sylla furnamed himfelfe *Fælix* ,not *Magnus, So Cæfar* faide to the Maifter of the fhippe, *Cæfarem portas & fortunam eius.*

But yet neuerthelefle thefe Pofitions *Faber quifq; fortunæ fuæ , fapiens dominabitur aftris : Inuia virtuti nu"aeft via,* and the like, being takē and vfed as fpurs to Induftry, and not as ftirops to in folency rather for i folution then for the prefumption or outwarde de-claration, haue beene euer thoughte founde and good, and are no queftion imprinted in the greateft mindes: who are fo fenfible of this opinion , as they cau fcarce containe t withi n. As we fee in *Auguſtus Cæfar*) who was rather diuerfe from his vncle, then, inferiour in vertue) how when he died, he defiered his friends aboute him to giue him a *Plaudite:* as if hee were confciente to himfelfe that he had played his parte wel vpon the ftage · This parte of knowledge we doe reporte alfo as deficient: not but that it is practifed too much, but it hath not beene reduced to writinge . And therefore leaft it fhoulde feme to any that it is not comprehenfible by Axiome, It is requi-fite as wee did in the former, that wee fet down fome heads or paffages of it.

Faber Fortu-næ siue de Am. bitu vitæ.

Wherein it maye appeare at the firft a newe and vnwoonted Argumente to teach men how to raife and make theire fortune a doctrine wherein euery man perchance will bee ready to yeeld him-felfe a difciple til he feethe difficulty: for *fortune* la-yeth as heauy impofitions as *vertue,* and it is as harde and fevere a thinge to bee a true *Polli-tique politique,* as to be truelye *morall.* But the handlinge hereof, concerneth learning greatly, both in honour, and in substance : In honour, becaufe pragmaticall men may not goe away with an opinion that learning is like a Larke that can mount, and singe, and pleafe her felfe, and nothing elfe; but may knowe that she houldeth as well of the hauke that can soare aloft, and can alfo defcend and strike vpon the pray. In substance, becaufe it is the perfite lawe of enquiry of trueth , *That nothing bee in the globe of matter, which should not be likewife in the globe of Cryftall, or Fourme,* that is, that there be uot any thing in being & action, which should not bee drawne and collected into contemplation and doctrine: Neyther doth learning admire or esteeme of this Architecture of fortune, otherwife then as of an inferiour worke; For no mans fortune can be an end woorthy of his being, and many times the woorthieft men doe abandon theyr fortune willingly for better refpects: but neuerthelesse fortune as an organ of vertue and merit deferueth the confideration.

First therefore the precept which I conceiue to bee moft fummary, towardes the preuayling in fortune; is to obtaine that windowe which *Momus* did require, who feeiug in the frame of mans heart, fuch Angles and receffes, founde fault there was not a windowe to looke into them : that is, to procure good informacions of particulars touching perfons, their Natures, their defires & ends, their cuftoms

and fashions, theyr helpes and aduantages, and wher-
by they cheefly stand; so againe their weaknesses and
disaduantages, and where they lye most open and obnoxious, their friendes, factions, dependaunces: and
againe theyr opposites, enuiors, competitors, theyr
moods, and times, *Sola viri molles adytus, & tēpora nor∝*
theyr principles, rules, and obseruacions: and the like;
And this not onely of persons, but of actions : what
are on foote from time to time: and how they are con
ducted, fauoured, opposed; and how they importe∗
and the like ; For the knowledge of present Actions, is not onely materiall in it selfe, but without it
also, the knowledge of persons is very erronious: for
men chaunge with the actions; and whiles they are
in pursuite, they are one , aud when they retourne
to theyr Nature, they are another . These Infor-
mations of particulars, touching persons and actions, are as the *minor* propositions in euery actiue syl
logisme, for no excellencye of obseruacions (which
are as the *maior* propositions) can suffice to ground a
conclusion, if there be error and mistaking in the mi-
nors.

That this knowledge is possible , *Salomon* is our
surety who sayeth. *Consilium in corde viri tanquam a-
qua profunda , sed vir prudens exhauriet illud* : And
although the knowledge it selfe falleth not vnder precept, because it is of Indiuiduals , yet the Instructions for the obtaining of it may.

We will beginne therefore with this precept, according to the aunciente opinion, that the Synewes

of wifedome, are flowneffe of beleefe, and diftruft : That more truft bee giuen to Countenances and Deedes,then to wordes : and in wordes, rather to fuddaine paffages , and furprifed wordes : then to fet and purpofed wordes: Neither let that be feared which is fayde,*fronti nulla fides*, which is meant of a generall outward behauiour,and not of the priuate and fubtile mocions and labours of the countenance and gefture,which as *Q. Cicero* elegantly fayth,is *A • nimi Ianua,the gate of the Mynd:*None more clofe then *Tyberius*,and yet *Tacitus* fayth of *Gallus*, *Etenim vultu offenfionem conieĉauerat.*So againe noting the differing Charaĉter and manner of his commending *Germanicus* and *Drufus* in the *Senate:*he fayeth, touching his fafhion wherein hee carried his fpeeche of Germnnicus,thus :*Magis in fpeciem adornatis verbis, quam vt penitus fentire crederetur* , but of Drufus thus, *Paucioribus fed intentior, & fida oratione :* and in another place fpeaking of his charaĉter of fpeech, when he did any thing that was gratious and populer,he fayeth,That in other thinges hee was *velut eluĉantium verborum:*but then againe , *Solutius loquebatur quando fubueniret.*So that there is no fuch artificer of diffimnlation : nor noe fuch commaunded countenaunce(*vultus iuffus*),that can feuer from a fained tale,fome of thefe fafhions, either a more fleight and careleffe fafhion,or more fet & formall, or more tedious and wandring: or comming from,a mã more drily and hardly.

Neither are *Deedes* fuch affured pledges,as that
Bbb 2   they

they may be trusted without a iudicious cōsideraciō of their magnitude and nature; *Fraus sibi in paruis fidem præstruit, vt maiore emolumento fallat* : and the Italian thinketh himselfe vpon the point to be bought and sould : when he is better vsed then he was woont to be without manifest cause. For small fauoures, they doe but lull men a sleepe, both as to *Caution*, and as to *Industry*, and are as *Demosthenes* calleth them, *Alimenta socordiæ*. So againe we see, how false the nature of some *deeds* are in that particular, which *Mutianus* practised vpon *Antonius Primus*, vpon that hollowe and vnfaithfull reconcilement, which was made betweene them: whereupon *Mutianus* aduanced many of the friends of *Antonius: Simul amicis eius præfecturas & tribunatus largitur:* wherein vnder pretence to strengthen him, he did desolate him, and won from him his dependances.

As for *words* (though they be like waters to Phisitions, ful of flattery and vncertainty) yet they are not to be dispised, specially with the aduantage of passion and affection . For so wee see *Tyberius* vpon a stinging and incensing speech of *Agrippina*, came a step foorth of his dissimulacion when he sayd, *You are hurt, because you doe not raigne:* of which *Tacitus* sayeth, *Audita hæc , raram occulti pectoris vocem elicuere correptamque Græco versu admonuit : ideo lædi quia non regnaret.* And therefore the Poet doth elegantly cal passions, tortures, that vrge men to confesse theyr secrets.

Vino tortus & ira.

And

And experience sheweth, there are few men so true to themselues , and so setled; but that sometimes vpon heate, sometimes vpon brauerye,sometime: vpon kindenesse, sometimes vpon trouble of minde and weaknesse,they open themselues;specially if they be put to it with a counter-dissimulatiõ,according to the prouerb of Spain,*Di mentira,y sacar as verdad:Tell a lye,and find a truth.*

As for the knowing of men, which is at second hand from Reportes : mens weakenesse and faultes are best knowne from theyr Enemies , theyr vertues, and abilityes from theyr friendes ; theyr customes aud Times from theyr seruauntes: their conceites and opinions from theyr familiar friends,with whom they discourse most:Generall fame is light,& the opinions conceiued by superiors or equals are deceitful:for to such men are more masked,*Verior fama è domesticis emanat.*

But the soundest disclosing and expounding of men is , by theyr natures and endes , wherein the weakest sorte of men are best interpreted by theyr Natures , and the wisest by theyr endes. For it was both pleasauntlye and wiselye sayde(though I thinke verye vntruely) by a Nuntio of the pope, returning from a certayne Nation , where hee serued as LIDGER : whose opinion beeing asked touching the appointemente of one to goe in his place, hee wished that in anye case they did not send one that was too wise : becaufe no very wise man would euer imagine,what they in that country
were

were like to doe : And certaynelye, it is an er_
rour frequent,for men to fhoot ouer,and to fuppofe
deeper ends,and more compaffe reaches then are :
the Italian prouerb being elegant,& for the moft part
true.

> *Di danari,di fenno,e di fede,*
> *C'n è manco che non credi:*

There is commonly leffe mony.leffe wifedome , and
leffe good faith,then men doe accompt vpon:
But Princes vpon a farre other Reafon are beft inter=
preted by their natures, and priuate perfons by theyr
ends,For Princes beeing at the toppe of humane de=
fires,they haue for the moft part no particular endes,
whereto they afpire : by diftaunce from which a man
mought take meafure and fcale of the reft of theyr
actions and defires . which is one of the caufes that
maketh theyr heartes more infcrutable: Neyther is
it fufficient to infourme onr felues in mens endes
and natures of the variety of them onely,but alfo of
the predominancy what humour reigneth moft,and
what end is principally fought.For fo wee fee, when
Tigellinus fawe himfelfe out_ftripped by *Petrouius*
Turpilianus in Neroes humours of pleafures *Metus*
eius rinaeur,he wrought vpon Neroes fears,wherbyhe
brake the others neck.

But to all this parte of Enquierie, the moft com_
pendious waye refteth in three thinges : The firft to
haue generall acquaintaunce and inwardneffe with
thofe which haue generall atquaintance, and looke
moft into the worlde : and fpeciallye according to

the

the diuerfitie of Bufineffe, and the diuerfitye of Perfons, to haue priuacye and conuerfation with fome
one friend at leaft which is parfite and well intelli-
genced in euery feuerall kinde. The feconde is to
keepe a good mediocritye in libertie of fpeeche, and
fecrecy in moft thinges libertye: fecrecy where it importeth: for libertye of fpeeche inuiteth and prouo-
keth libertye to bee vfed againe : and fo bringeth
much to a mans knowledge: and fecrecie on the o-
ther fide induceth truft and inwardneffe. The laft is
the reducing of a mans felfe to this watchfull and
ferene habite, as to make accompte and purpofe in e-
uerye conference and action, afwell to obferue as to
acte. For as *Epictetus* would haue a Philofopher in eue
ry particular action to fay to himfelfe, *Et hoc volo, &*
etiam inftitutum feruare: fo a politique man in euerye
thing fhould fay to himfelf; *Et hoc volo, ac etiam aliquid*
addifcere. I haue ftaied the longer vpon thit precept of
obtaining good information, becaufe it is a maine
part by it felfe , which aunfwereth to all the reft.
But aboue al things, Caution muft be taken, that Men
haue a good ftaye and houlde of themfelues, and that
this much knowing doe not draw on much medling:
For nothing is more vnfortunate then light and rafh
intermeddling in many matters: So that this v⁻ ietyof
knowledge tendeth in conclufion but onely to this, to
make a better & freer choife of thofe actions, which
may concern vs, & to conduct them with the leffe error and the more dexterity.
The fecond precept concerning this knowledge is
for.

for men to take good informacion touching theyre own perfon and well to vnderftand themfelues:know, ing that as S. *Iames* fayth , though men looke oft in a glaffe,yet they do fodainly forget themfelues, wher- in as the diuine glaffe is the word of God, fo the po- litique glaffe is the ftate of the world, or times wher- in we liue:In the which we are to behould our felues.

For men ought to take an vnpartiall viewe of their owne abilities and vertues:and againe of their wants and impediments;accounting thefe with the moft, & thofe other with the leaft,and from this view and exa- mination to frame the confiderations following. Firft to confider how the conftitution of their nature forteth with the generall ftate of the times : which if they find agreeable and fit, then in all things to giue themfelues more fcope and liberty,but if differing and diffonant,then in the whole courfe of theyr life to be more clofe retyred and referued:as we fee in *Tyberius* who was neuer feen at a play: and came not into the Senate in 12.of his laft yeers: whereas *Auguftus Cafar* liued euer in mens eyes, which *Tacitus* obferueth, *Alia Tiberio morum via.*

Secondly to confider how their Nature forteth with profeffions and courfes of life,& accordingly to make election if they be free,and if ingaged,to make the de parture at the firft opportunity: as we fee was doone by Duke *Valentine*,that was defigned by his father to a facerdo tal profeffion,but quitted it foon after in re- gard of his parts and inclination being fuch neuerthe- leffe,as a man cannot tel wel whether they were worfe

for

for a Prince or for a Prieſt.

Thirdly ro conſider how they ſorte with thoſe whom they are like to haue Competitors and Con-currents and to take that courſe wherin there is moſt ſolitude, and themſelues like to be moſt eminent : as *Cæſar Iulius* did, who at firſt was an Orator or Pleader but when he ſaw the excellency of *Cicero*, *Hortenſi-us*, *Catulus*, and others for eloquence, and ſawe there was no man of reputation for the warres but *Pompeius* vpon whom the State was forced to relie ; he for-ſooke his courſe begun toward a ciuile and popular greatneſſe ; and tranſferred his deſignes to a mar-ſhall greatneſſe.

Fourthly in the choyſe of their friends, and de-pendaunces, to proceeed according to the Compo-ſiticn of their own nature, as we may ſee in *Cæſar*, all whoſe friends and followers were men actiue and ef-fectuall, but not ſolemn or of reputation.

Fiftly to take ſpeciall heed how they guide them ſelues by examples, in thinking they can doe as they ſee others doe :whereas perhappes their natures and carriages are farre differing; in which Error, it ſee-meth *Pompey* was, of whome *Cicero* ſayeth, that hee was woont often to ſaye : *Sylla potuit*; *Ego non potero?* wherein he was much abuſed, the natures and pro-ceedinges of himſelfe and his example , beeing the vnlikeſt in the worlde, the one being fierce, vi-olent, and preſſing the fact; the other ſolemn, and full of Maieſty and circumſtance, and therefore the leſſe effectuall.

But this precept touching the politicke knowledge of our selues hath many other branches whereupon we cannot infist:

Next to the wellvnderftanding and difcerning of a mans felfe, there followeth the well opening and reuealing a mans felfe, wherein we fee nothing more vfuall then for the more able man to make the leffe fhewe. For there is a greate aduantage in the well fetting foorthe of a mans vertues, fortunes, merites, and againe in the artificiall couering of a mans weakeneffes, defectes, difgraces, ftaying vpon the one flyding from the other, cherifhing the one by circumftaunces, gracing the other by expofition, and the like; wherein we fee what *Tacitus* fayth of *Mutianus*, who was the greateft politique of his time, *Omnium quæ dixerat fecerat que, arte quadam oftentator:* which requireth indeed fome arte, leaft it turne tedious and arrogant, but yet fo;as oftentation (though it be to the firft degree of vanity) feemeth to me rather a vice in Manners, then in Policye; for as it is fayd, *Audacter calumniari, femper aliquid hæret,* So except it be in a ridiculous degree of deformity *Audacter te vendita femper aliquid hæret.* For it will fticke with the more ignoraunt and inferiour fort of men, though men of wifedome and ranke doe fmile at it and defpife it, and yet the authority wonne with many, doth counteruaile the difdaine of a few. But if it be carried with decency and gouernement, as with a naturall pleafaunt and ingenious fafhion, or at times when it is mixte with fome perill and vnsafety, (as in Military perfons) or at tymes when others are moſt enuied; or with eaſie and careleſſe paſſage to it and from it, without dwelling too long, or being too ſerious: or with an equall freedome of taxing a mans ſelfe, aſwell as gracing himſelfe, or by occaſion of repelling or putting downe others iniurye or inſolencie : It doth greately adde to reputation; and ſurelye not a fewe ſolide natures, that wante this ventoſitye, and cannot ſaile in the heighth of the windes, are not without ſome preiudice and diſaduantage by theyre moderacion.

But for theſe flouriſhes and inhanſements of vertue, as they are not perchaunce vnneceſſary : So it is at leaſte, neceſſary that vertue be not diſualewed and imbaſed vnder the iuſt price : which is doon in three manners; By offering and obtruding a mans ſelfe; wherein men thinke he is rewarded when he is accepted. By doing too much, which wil not giue that which is well done leaue to ſettle, and in the end induceth ſaciety:and: By finding to ſoone the fruit of a mans vertue, in commendation, applauſe, honour, fauoure, wherein if a man be pleaſed with a little, let him heare what is truly ſaid, *Caue ne inſuetus rebus maioribus videaris, ſi hæc te res parua ſicuti magna delectat :*

But the couering of defects is of no leſſe importance, then the valewing of good parts. which may be doone likewiſe in three manners, by *Caution,* by *Colour,* and by *Confidence.* *Caution* is, when men doe

ingenioufly and difcreetely auoyde to be put into
thofe things for which they are not propper:wher-
as contrarywife bould and vnquiet fpirits will thruft
themfelues into matters without difference,and fo
publifh and proclaime all their wantes; *Coloure* is
when men make a way for themfelues, to haue a
conftruction made of their faultes or wantes : as
proceedinge from a better caufe, or intended for
fome other purpofe : for of the one, it is well
fayde;

Sæpe latet vitium proximitate boni:
And therefore whatfoeuer want a man hath, he muft
fee, that he pretend the vertue that fhadoweth it,
as if he be *Dull*,he muft affect *Grauitie*, if a *Cowarde*,
Mildeneffe, and fo the reft : for the fecond, a man
muft frame fome probable caufe why he fhould not
doe his beft,and why he fhould diffemble his abilities : and for that purpofe muft vfe to diffemble
thofe abilities, which are notorious in him to giue
colour that his true wants are but induftries and dif-
fimulations : for *Confidence* it is the laft but the fu-
reft remedie : namely to depreffe and feeme to defpife whatfoeuer a man cannot attaine, ooferuing the
good principle of the Marchantes, who endeuour
to raife the price of their owne commodities,and to
beate down the price of others. But there is a confidence that paffeth this other : which is to face out a
mans own defects: in feeming to conceiue that he is
beft in thofe things wherein he is failing : and to help
that againe, to feeme on the other fide that he hath
leaft

leaft opinion of himfelfe, in thofe things wherein he is beft: like as we fhall fee it commonly in Poets, that if they fhew their verfes, and you except to any, they will fay, *That that lyne coſt them more labour then any of the reſt* : and prefently will feeme to difable, and fufpect rather *ſome other lyne*, which they know well enough to be the beft in the number. But aboue al in this righting and helping of a mans felfe in his owne carriage, he muft take heed he fhew not himfelfe difmantelled and expofed to fcorne and iniury, by too much dulceneffe, goodneffe, and facility of nature, but fhew fome fparkles of liberty, fpirit, and edge. Which kind of fortified cariage with a readye refcuffing of a mans felfe from fcornes, is fomtimes of neceffity impofed vpon men by fomwhat in their perfon or fortune, but it euer fucceedeth with good felicity.

Another precept of this knowledge is by all poffible endeauour, to frame the mind to be pliaunt and obedient to occafion; for nothing hindereth mens fortunes fo much as this : *Idem manebat, nequeidem decebat :* Men are where they were, when occafions turne, and therefore to *Cato*, whom *Liuie* maketh fuch an Architect of fortune, hee addeth that he had *Verſatile Ingenium :* And thereof it commeth that thefe graue folemne wittes which muft be like themfelues, and cannot make departures haue more dignity then fœlicity : But in fome it is nature to bee fomewhat vifcoufe and inwrapped,

and

and not eafie to turne : In fome it is a conceipte, that is almoft a nature, which is that men can hardlie make themfeluns beleeue that they oughte to chaunge their courfe, when they haue found good by it in former experience. For *Macciaue* noteth wifely how *Fabius Maximus* would haue been temporizing ftill, according to his ould biaffe, when the nature of the warre was altered, and required hotte purfuite; In fome other it is want of point and penetration in their iudgemente, that they do not difcerne when thinges haue a periode, but come in, too late after the occafion : As *Demofthenes* compareth the people of *Athens* to country fellowes, when they play in a fence fchoole, that if they haue a blow then they remooue their weapon to that warde, and not before : In fome other it is a loathneffe to leefe labours paffed, and a conceite that they can bring about occafions to their plie, and yet in the end, when they fee no other remedye, then they come to it with difaduantage, as *Tarquinius* that gaue for the third part of *Sybillaes* bookes the treeble price, when he mought at firft haue had all three for the fimple. But from whatfoeuer roote or caufe this Reftiueneffe of mind proceedeth, it is a thing moft preiudiciall, and nothing is more politique then to make the wheels of our mind concentrique and voluble with the wheels of fortune.

Another precept of this knowledge, which hath fome affi nity with that vve laft fpake of, but with difference is that which is well expreffed, *fatis accede*

Deif-

Deisque, that men do not only turne with the occasions but also runne with the occasions and not strain their credit or strength to ouer harde or extreame points : but choose in their actions that which is most passable : for this will preserue men from foyle, not occupy them too much about one matter, winne opinion of moderation , please the moste , and make a showe of a perpetuall fœlicitye in all they vndertake , which cannot but mightely increase reputation.

Another part of this knowledge seemeth to haue some repugnancy with the former two, but not as I vnderstand it, and it is that which *Demosthenes* vttereth in high tearmes: *Et quemadmodum receptum est, vt exercitum ducat Imperator: sic & a cordatis viris res ipsæ ducendæ, vt quæ ipsis videntur, ea gerantur, & non ipsi euentus persequi cogantur.* For if we obserue, we shall find two differing kinds of sufficiency, in managing of businesse : some can make vse of occasions aptly and dexteroufly, but plotte little: some can vrge and pursue their owne plottes well, but cannot accommodate nor take in: either of which is very vnperfite without the other.

Another part of this knowledge is the obseruing a good mediocrity in the declaring or not declaring a mans selfe, for although depth of secrecy, and making way (*qualis est via nauis in Mari* , which the French calleth *Sourdes Menees*, when men set thinges in worke without opening themselues at all) be somtimes

times both profperous and admirable : yet many times *Diffimulatio errores parit, qni diffimulatorem ip-fum illaqueant.* And therefore we fee the greateft pol-litiques haue in a naturall and free manner profeffed their defires, rather then bin referued and difguifed in them. For fo we fee that *Lucius Sylla* made a kind of profeffion, *That hee wifhed all men happy or vnhappie as they ftood his friendes or enemies* : So *Cæfar*, when hee went firft into *Gaul*, made no fcruple to profeffe, *that hee had rather bee firft in a village, then fecond at Rome.* So againe as foone as hee had begunne the warre, we fee what *Cicero* fayth of him, *Alter* (mea-ning of *Cæfar*) *non recufat, fed quodam modo poftulat, vt* (*vt eft*) *fic appelletur Tyrannus.* So we may fee in a letter of *Cicero* to *Atticus*, that *Auguftus Cæfar* in his very entrance into affaires, when he was a dearling of the Senate, yet in his haranges to the people, would fweare *Ita parentis honores confequi liceat*, (which was no leffe then the Tyranny,) faue that to helpe it, hee would ftretch foorth his hand towardes a ftatua of *Cæfars*, that was erected in the place : and men laughed and woondered and fayde, Is it pof-fible, or did you euer heare the like, and yet though hee meant no hurte, hee did it fo hand-femlye and ingenuouflye, and all thefe were prof-perous, where as *Pompeye* who tended to the fame ends, but in a more darke and diffembling manner, as *Tacitus* fayeth of him, *Occultior non melior,* wherein *Saluft* concurreth *ore probo, animo inuerecun-*
do,

do made it his disseigne by infinite secret Engines, to caste the state into an absolute Anarchy and confusion, that the state mought cast it selfe into his Armes for necessity and protection, and so the soueraigne power bee putt vpon him, and he neuer seene in it : and when hee had broughte it (as he thoughte)to that pointe when hee was chosen *Consull* alone; as neuer any , was; yet hee could make noe greate matter of it, becaufe men vnderstoode him not: but was faine in the end, to goe the beaten tracke of getting Armes into his handes, by coulour of of the doubte of *Cæsars* dessignes: so tedious, casuall, and vnfortunate are thefe deepe dissimulations, whereof it seemeth *Tacitus* made this iudgement, that they were a cunning of an inferiour fourme in regard of true pollicy, attributing the one to *Augustus*, the other to *Tiberius*, where speaking of *Liuia*, he sayth: *Et cum artibus mariti simulatione filij bene composita:* for furely the continuall habite of dissimulation is but a weake and fluggish cunning, & not greatly politique.

Another precept of this Architecture of Fortune, is, to accuftome our mindes to iudge of the proportion or valewe of things, as they conduce, and are materiall to our particular ends, and that to doe fubftantially and nor fuperficially ⟡ For wee fhall finde the Logicall parte (as I maye tearme it) of fome mens mindes good, but the Mathematicall part erroneous, that is, they can well iudge of confequences , but not of proportions and

com

comparison, preferring things of shewe and sence before things of substance and effect. So some fall in loue with accesse to Princes, others with popular fame and applause, supposinge they are things of greate purchase, when in many Cases they are but matters of Enuy, perill, and Impediment: So some measure thinges accordinge to the labour and difficulty, or assiduity, which are spent aboute them; and thinke if they bee euer mcovinge, that they must needs aduance and proceede, as *Cæsar* saith in a dispisinge manner of Cato the second, when hee describeth howe laborious and indefatigable he was to noe greate purpose: *Hæc omnia magno studio agebat.* So in moste thinges men are ready to abuse themselues in thinking the greatest means to be best, when it should bee the Fittest.

As for the true marshalling of mens pursutes towards theire fortune as they are more or lesse materiall, I houlde them to stand thus; Firste the amendment of their own Minds. For the Remooue of the Impediments of the mind wil sooner cleare the passages of fortune, then the obteininge fortune wil remooe the Impedimēts of the mind; In secōd place I set downe wealth and meanes, which I know most men woulde haue placed firste: because of the generall vse which it beareth towardes all varietie of occasions. But that opinion I may condemne with like reason as *Macchiauell* doth that other: that monies were the sinews of the warres, wheras (saith he)

the true finews of the warres are the finews of mens
Armes, that is a valiant, populous and Military
Nation:& he voucheth aptly the authority of *Solon*
who when *Crœfus* fhewed him his treafury of goulde
faide to him, that if another came that had better
Iron, he woulde be maifter of his Gould, In like manner it may be truly affirmed, that it is not menies that
are the finews of fortune, but it is the finews and
fteele of mens Mynds, VVitte, Courage, Audacity,
Refolution, Temper, Induftry, and the like: In thirde
place I fet down Reputation, becaufe of the peremptory Tides & Currants it hath, which if they bee not
taken in their due time, are fildome recouered, it
beinge extreame harde to plaie an after game
of reputation. And laftly, I place honoure, which
is more eafily wonne by any of the other three,
much more by all, then any of them can bee pur-
chafed by honour. To conclude this precepte, as
there is order and priority in Matter, fo is there
in Time, the propofterous placing whereof is one
of the commoneft Errors:while men fly to their ends
when they fhoulde intend their beginninings: and
doe not take things in order of time as they come on,
but marfhall them according to greatnes and not
according to inftance, not obferuing the good precepte *Quod nunc inftat agamus.*

Another precept of this knowledge is, not
to imbrace any matters, which doe occcupie to great
a quantity of time, but to haue that founding in a
mans eares.

Sed

Sed fugit interea, fugit irreparabile tempus, and that is the caufe why thofe which take their courfe of ri-fing by profeſſions of Burden, as Lawyers, Orators painefull diuines, and the like, are not common-lie fo politique for their owne fortune, otherwife then in their ordinary way, becaufe they want time to learne particulars, to waite occafions, and to deuife plottes.

Another precept of this knowledge is to imitate nature which doth nothing in vaine, which furely a man may do, if he do well interlace his bufineſſe, and bend not his mind too much vpon that which he prin cipally intendeth. For a man ought in euery particu-lar action, fo to carry the motions of his mind, and fo to haue one thing vnder another, as if he cannot haue that he feeketh in the beſt degree, yet to haue it in a fecond, or fo in a third, and if he can haue no parte of that which he purpofed, yet to turn the vfe of it to fo-what els, and if he cannot make any thing of it for the prefent, yet to make it as a feed of fomwhat in time to come, and if he can contriue no effect or fubftaunce from it, yet to win fom good opinion by it, or the like fo that he fhould exact an account of himfelf of eue-ry action, to reape fomwhat, and not to ſtand amazed and confufed if he faile of that he chiefly meant : for nothing is more impollitique then to mind actions wholly one by one. For he that dooth fo , leefeth infinite occafions which enterueine, and are many times more proper and propitious for fomewhat, that he fhall need afterwards : then for that which

he

he vrgeth for the present; and therfore men must be
parsite in that rule: *Hæc oportet facere, & illa non os*
mittere.

Another precept of this knowledge is, not to in-
gage a mans selfe peremptorily in any thing, though
it seem not liable to accident, but euer to haue a win-
dow to flie out at, or a way to retyre; following the
wisedom in the ancient fable, of the two frogs, which
consulted when their plash was drie , whether they
should go:and the one mooued to go down into a pit
because it was not likely the water would dry there,
but the other answered, *True, but if it do, how shall we get*
out againe?

Another precept of this knowledge is that ancient
precept of *Bias*, construed not to any point of perfidi-
ousnesse, but only to caution and moderation *Et ama*
tanquam inimicus futurus, & odi tanquam amaturus: For
it vtterly betraieth al vtility, for mē to imbarque them
selues to far, into vnfortunate friendships:troublesom
spleans;& childish & humorous enuies or æmulatiōs.

But I continue this beyond the measure of an example, led, because I wold not haue such knowledges
which I note as *deficient* to be thought things Imaginatiue, or in the ayre; or an obseruation or two, much
made of but thinges of bulke and masse: whereof an
end is hardlier made, then a beginning. It must be
likewise conceiued that in these pointes which I
mencion and set downe, they are far from complete
tractates of them : but onelye as small peeces for
patternes : And lastlye , no man I suppose will
thinke

thinke, that I meane fortunes are not ob_
teyned without all this adoe; For I know they come
tumblinge into some mens lappes, and a nomber
obtaine good fortunes by dilligence, in a plaine way:
Little intermedlinge: and keeping themselues from
grosse errors.

But as Cicero when he setteth down an *Idea* of a par-
fit Orator, doth not mean that euery pleader should
be such; and so likewise, when a Prince or a *Courtier*
hath been described by such as haue handled those
subiects, the mould hath vsed to be made accordinge
to the perfectiõ of the Arte, and not according to cõ
mon practise : So I vnderstand it that it ought to be
done in the description of a *Pollitique* man. I meane
pollitique for his owne fortune.

But it must be remembred al this while, that the precepts which we haue set down, are of thatkind which
may be coũted & called *Bonæ Artes*, as for euill arts, if
a man would set down for himselfe that principle of
*Machiauel: That a man seeke not to attaine vertue it
selfe: But the apparance onely thereof, because the credite of
vertue is a helpe, but the vse of it is cumber*: or that o-
ther of his principles: *That he presuppose, that men are
not fitly to be wrought otherwise but by feare, and therefore
that he seeke to haue euery mã obnoxius, lowe, & in streight*
which the *Italiãs* cal *seminar spine*, to sowe thornes: or
that other principle cõteined in the verse which *Cice
ro* cyteth *cadant amici, dũmodo Inimici intercidãt, as the
Trium virs* which sould euery one to other the liues
of their friends for the deaths of theire enemiees: or
that

that other protestation of *L. Catilina* to set on fire &
trouble states, to the end to fish in droumy waters, &
to vnwrappe their fortunes. *Ego si quid in fortunis meis*
excitatum sit incendium, id non aqua sed ruina restinguam,
or that other principle *of Lysäder That childrē are to be*
deceiued with cōfittes, & men with othes, & the like euil
and corrupt positions, whereof (as in al things) there
are more in number then of the good: Certainly with
these dispensations from the lawes of charity & inte_
gryty the pressing of a mans fortune, may be more ha
sty and compendious. But it is in life, as it is in ways
The shortest way is comonly the fowlest: & surely the
fairer way is not much about.

But men if they be in their own power & doe beare &
sustaine themselues, and bee not caryed awaye with
a whirle winde or tempest of ambition: oughte in the
pursute of their owne fortune, to set before their eies,
not only that general Map of the world. *That al things*
are vanity & vexatiō of spirit, but many other more par
ticular Cards & directiōs, cheefly that, That Being,
without wel being: is a curse; & the greater being, the
greater curse, And that all vertue is most rewarded, &
al wickednesse most punished in it selfe: according as
the *Poet* saith excellently.

   Quæ vobis que digna viri, pro laudibus istis
   Premia posse rear solui? pulcherrima primum
    Dij moresque *dabunt vestri:*

And so of the contrary. And secondly they oughte to
looke vp to the eternal prouidence and diuine iudgemente, which often subuerteth the wisdome of euyll
                         plot

plots & imaginations, according to that scripture *He hath conceiued mischiefe & shal bring foorth a vainething* And although men should refraine themselues from iniury and euil artes, yet this incessant & Sabbathlesse pursute of a mans fortune, leaueth not tribute which we owe to God of our time who (we see) demandeth a tenth of our substāce, & a seauenth, which is more strict of our time: and it is is to smal purpose to haue an erected face towards heauē, & a perpetual groueling spirit vpon earth eating dust as doth the serpent, *Atque affigit humo Diuinæ particulam auræ:* And if any mā flatter himself that he will imploy his fortune wel, though he shold obtain it ill, as was said concerning *Aug. Cæsar*, & after of *Septimius Seuerus, That either they shold neuer haue bin born or else they shold neuer haue died,* they did so much mischief in the pursut & ascētof their greatnes, & so much good when they were established yet these cōpensations & satisfactions, are good to be vsed, but neuer good to be purposed: And lastly it is not amisse for mē in their race toward their fortune to cooll thēselues a litle with that cōceit which is elegātly expressed bythe Emperor, *Charls* the 5. in his instructiōs to the K. his son, *That fortune hath sōwhat of the nature of a womā, that if she be too much woed, she is the farder of.* But this last is but a remedy for those whose Tasts are corrupted: let mē rather build vpo that foūdation which is as a cornerstone of diuinity and philosophie, wherein they ioyne close, namely that same *Primum quærite.* For diuinity sayth *Primum quærite regnū Dei & ista omnia adiiciētur Vobis:* & Philosophy saith, *Primū quærite bona animi, cœtera aut aderunt, aut non oberunt.*

And

And although the humane foundation hath somewhat of the same, as we see in M: *Brutus* when hee
brake forth into that speech.

   - - Te colui(Virtus)vt rem: aſt tu nomen inane es;
Yet the diuine foundation is vpon the Rocke. But
this may serue for a Taſt of that knowledge which I
noted as deficient.

   Concerning gouernment, it is a part of knowledge, secret and retyred in both these respects, in
which things are deemed secret: for some things are
secret, becauſe they are hard to know, and some becauſe they are not fit to vtter : wee see all gouernments are obſcure and inuisible.

   - - - Totamque infuſa per artus,
   Mens agitat molem, & magno corpore miſcet.
Such is the description of gouernments ; we see the
gouernmēt of God ouer the world is hidden; inſomuch as it seemeth to participate of much irregula-
ritie and confuſion; The gouernment of the Soule
in moouing the Body is inward and profound, and
the paſſages therof hardly to be reduced to demonſtration. Againe, the wiſedome of Antiquitie (the
ſhadowes whereof are in the Poets) in the deſcripti-
on of torments and paines, next vnto the crime of
Rebellion, which was the Giants offence, doth de-
teſt the offence of facilitie : as in *Syſiphus* and *Tanta-*
lus. But this was meant of particulars; Neuertheleſſe
euen vnto the generall rules and diſcourſes of pollicie, and gouernment, there is due a reuerent and reſerued handling.

         E e e   But

But contrariwise in the gouernors towards the gouerned, all things ought as far as the frailtie of Man
permitteth, to be manifeft, &reuealed. For fo it is expreffed in the Scriptures touching the gouernment
of God, that this Globe which feemeth to vs a dark
and fhady body is in the view of God, as Chriftall,
Et in confpectu fedis tanquã mare vitreũ fimile chriftallo.
So vnto Princes and States, and fpecially towardes
wife Senats and Councels, the natures and difpofitions of the people, their conditions, and neceffi-
ties, their factions and combinations, their animofi-
ties and difcontents ought to be in regard of the varietie of their Intelligences, the wifedome of their
obferuations, and the height of their ftation, where
they keepe Centinell, in great part cleare and tran-
fparent ; wherefore, confidering that I write to a
king that is a maifter of this Science, and is fo wel af-
fifted, I thinke it decent to paffe ouer this part in fi-
lẽce, as willing to obtaine the certificate, which one
of the ancient Philofophers afpired vnto, who being
filent, when others contended to make demonftration of their abilities by fpeech, defired it mought
be certified for his part, *that there was one that knewe
how to hold his peace.*

Notwithftanding for the more publique part of
Gouernment, which is Lawes, I think good to note
onley one deficience, which is, that all thofe which
haue writtẽ of Lawes, haue written either as Philofophers, or as lawiers, & none as Statesmen. As for
the Philofophers, they make imaginary Lawes for
imaginary cõmon-wealths, & their difcourfes are as
the

the Stars, which giue little light becaufe they are fo high. For the Lawyers, they write according to the States where they liue, what is receiued Law, & not what ought to be Law; For the wifedome of a Lawmaker is one, & of a Lawyer is another. For ther are in Nature certaine fountaines of Iuftice, whence all Ciuil Lawes are deriued, but as ftreames; & like as waters doe take tinctures and taftes from the foyles through which they run; So doe ciuill Lawes vary according to the Regions and gouernments where they are plâted, though they proceed from the fame fountaines; Againe the wifedome of a Lawmaker confifteth not onely in a platforme of Iuftice; but in the application thereof, taking into confideration, by what meanes Lawes may be made certaine, and what are the caufes & remedies of the doubtfulneffe and *incertaintie* of Law, by what meanes Lawes may be made apt and eafie to be executed, and what are the impediments, and remedies in the *execution* of lawes, what influence lawes touching priuate right of M*eum & Tuum*, haue into the publike ftate, and how they may be made apt and *agreable,* how lawes are to be *penned* and *deliuered,* whether in *Texts* or in *Acts, briefe* or *large,* with *preambles,* or *without* howe they are to bee *pruned* and *reformed* from time to time, and what is the beft meanes to keepe them frô being too *vaft in volumes,* or too ful of *multiplicitie* & *crofneffe,* how they are to be expounded, When *vpon caufes emergent,* and iudicially dif-cuffed, and when vpon *refponfes* and conferences touching generall points or queftions, how they

are to be *preſſed*, rigorouſly, or tenderly, how they
are to be *Mitigated* by equitie and good conſcience,
and whether diſcretion and ſtrict Lawe are to be
mingled in the ſame Courts, or *kept a part in ſeuerall
Courts*; Againe, how the practiſe, profeſſion, and e-
rudition of Lawe is to be cenſured and gouerned,
and many other points touching the adminiſtration, and (as I may tearme it) animation of Lawes.
Vpon which I inſiſt the leſſe, becauſe I purpoſe (if
God giue me leaue) hauing begunne a worke of
this Nature, in Aphoriſmes, to propound it hereafter, noting it in the meane time for deficient.

*De prudē-
tia legisla=
toria, ſiue,
de ſontibus
Iuris.*

And for your Maieſties Lawes of *England*, I
could ſay much of their dignitie, and ſomewhat of
their defect : But they cannot but excell the ciuill
Lawes in fitneſſe for the gouernment: for the ciuill
Law was, *non hos quæſitum munus in vſus*: It was not
made for the countries which it gouerneth : hereof
I ceaſe to ſpeake , becauſe I will not intermingle
matter of Action , with matter of generall Learning.

THus haue I concluded this portion of learning
touching *Ciuill knowledge*, & with Ciuill knowledge haue concluded HVMANE PHILO-
SOPHY and with Humane Philoſophy, PHI-
LOSOPHY in GENERAL; and being now
at ſome pauſe, looking backe into that I haue paſſed
through: This writing ſeemeth to me (*Si nunquam
fallit imago*) as farre as a man can iudge of his owne
worke,

worke, not much better then that noise or sound which Musitiãs make while they are in tuning their Instrumẽts, which is nothing pleasãt to hear, but yet is a cause why the Musique is sweeter afterwardes. So haue I beene content to tune the Instruments of the Muses, that they may play, that haue better hands. And surely when I set before me the condition of these times, in which learning hath made her third visitation, or circuite in all the qualities thereof: as the excellencie and viuacitie of the wits of this age; The noble helpes and lights which we haue by the trauailes of ancient writers: The Art of Printing, which communicateth Bookes to men of all fortunes. The opénesse of the world by Nauigation, which hath disclosed multitudes of experiments, and a Masse of Naturall History: The leasure wherwith these times abound, not imploying men so generally in ciuill businesse, as the States of *Græcia* did, in respect of their popularitie, and the State of *Rome* in respect of the greatnesse of their Monarchie : The present disposition of these times at this instant to peace: The consumption of all that euer can be said in controuersies of Religiõ, which haue so much diuerted men from other Sciences : The perfection of your Maj: learning, which as a *Phœnix* may call whole volyes of wits to followe you, and the inseparable proprietie of Time, which is euer more and more to disclose truth : I cannot but be raised to this perswasion, that this third period of time will farre surpasse that of the *Græcian* and *Romane*

mane Learning: Onely if men will know their own strength, and their owne weakeneſſe both: and take one from the other, light of inuention, and not fire of contradiction, and eſteeme of the Inquiſition of truth, as of an enterpriſe, & not as of a qualitie or ornament, & imploy wit and magnificence to things of worth & excellencie, & not to things vulgar, and of popular eſtimation. As for my labors, if any man ſhall pleaſe himſelfe, or others in the reprehenſion of them; they ſhall make that ancient and patient requeſt, *verbera, ſed audi.* Let men reprehend them ſo they obſerue and waigh them: For the Appeale is lawfull (though it may be it ſhall not be needefull) from the firſt cogitations of men to their ſecond, & from the neerer times, to the times further of. Now let vs come to that learning, which both the former times were not ſo bleſſed as to knowe, *Sacred & inſpired Diuinitie,* the Sabaoth and port of all mens labours and peregrinations.

The prerogatiue of God extendeth as well to the reaſon as to the will of Man; So that as we are to obey his law though we finde a reluctatió in our wil: So we are to belieue his word, though we finde a reluctation in our reaſon : For if we belieue onely that which is agreeable to our ſence, we giue conſent to the matter, and not to the Author, which is no more then we would doe towards a ſuſpected and diſcredited witneſſe : But that faith which was accounted to *Abraham* for righteouſneſſe

neſſe was of ſuch a point, as whereat *Sarah* laughed,
who therein was an Image of Naturall Reaſon.

Howbeit (if we will truly conſider of it) more
worthy it is to belieue , then to knowe as we now
know; For in knowledge mans mind ſuffereth from
ſence, but in beliefe it ſuffereth from Spirit, ſuch one
as it holdeth for more a uthoriſed then it ſelfe, & ſo
ſuffereth from the worthier Agent: otherwiſe it is of
the ſtate of man glorified, for then faith ſhal ceaſe, &
we ſhall knowe as we are knowne.

Wherefore we conclude that ſacred Theologie
(which in our Idiome we call Diuinitie) is grounded onely vpon the word & oracle of God, and not
vpon the light of nature: for it is written, *Cæli enarrāt
gloriam Dei*: But it is not written *Cæli enarrant volun-
tatem Dei*: But of that it is ſaid; *Ad legem & teſtimoni-
um ſi non fecerint ſecundū verbum iſtud &c.* This holdeth not onely in thoſe points of faith , which concerne the great miſteries of the Deitie, of the Creation, of the Redemption, but likewiſe thoſe which
concerne the law Moral truly interpreted; *Loue your
Enemies, doe good to thē that hate you. Be like to your heauenly father, that ſuffereth his raine to fal vpon the Iuſt &
Vniuſt.* To this it ought to be applauded, *Nec vox ho-
minē ſonat,* It is a voice beyond the light of Nature:
So we ſee the heathen Poets when they fall vpon a
libertine paſſion , doe ſtill expoſtulate with lawes
and Moralities , as if they were oppoſite and
malignant to Nature : *Et quod natura remittit,*
inuida

inuida Iura negant: So faid *Dendamis* the *Indian* vnto
Alexanders Meſſengers : That he had heard ſomewhat of *Pythagoras* ,and ſome other of the wiſe men
of *Græcia*, and that he held them for excellent Men:
but that they had a fault , which was that they had
in too great reuerence and veneration a thing they
called Lawe and Manners: So it muſt be confeſſed
that a great part of the Lawe Morall is of that perfection , whereunto the light of Nature cannot a-
ſpire: how then is it,that man is ſaide to haue by the
light and lawe of Nature ſome Notions, and conceits of vertue and vice, iuſtice & wrong, good and
euill? Thus, becauſe the light of Nature is vſed in
two ſeuerall ſenſes : The one,that which ſpringeth
from Reaſon, Senſe, Induction, Argument,accor-
ding to the lawes of heauen and earth : The other
that which is imprinted vpon the ſpirit of Man by
an inward Inſtinct,according to the lawe of conſcience , which is a ſparkle of the puritie of his firſt E-
ſtate: In which later ſenſe onely,he is participant of
ſome light,and diſcerning:touching the perfection
of the Morall lawe:but how? ſufficient to check the
vice,but not to informe the dutie. So then the doctrine of Religion,as well Morall as Miſticall, is not
to be attained , but by inſpiration and reuelation
from God.

    The vſe notwithſtanding of Reaſon in ſpirituall
things,and the latitude thereof is very great and ge-
nerall:for it is not for nothing, that the Apoſtle calleth Religion *our reaſonable ſeruice of God* , inſo-
                               much

much as the verie Ceremonies and Figures of the oulde Lawe were full of reason and signification, much more then the ceremonies of Idolatrie and Magicke, that are full of *Non-significants* and *Surde Characters*; But most specially the Christian faith, as in all things, so in this deserueth to be highly magnified, houlding and preseruing the golden Mediocritie in this point, betweene the law of the *Heathen*, and the law of *Mahumet*, which haue embraced the two extreames.   For the Religion of the *Heathen* had no constant beleefe or confession, but left all to the libertie of argument: and the Religion of *Mahumet* on the other side, interdicteth argument altogether; the one hauing the verie face of Errour : and the other of Imposture ;   whereas the Faith doth both admit and reiect Disputation with difference.

The vse of Humane Reason in Religion, is of two sorts : The former in the conception and apprehension of the Mysteries of G o d to vs reuealed ; The other, in the inferring and deriuing of doctrine and direction thervpon : The former extendeth to the mysteries themselues: but how? by way of Illustration, and not by way of argument. The later consisteth indeed of Probation and Argument.   In the former wee see God vouchsafeth to descend to our capacitie, in the expressing of his misteries in sort as may bee sensible vnto vs : and doth grifte his Reuelations & holie doctrine vpon the Notions of our reason, and applyeth his Inspi-

F f f   ration

ratiōs to open our vnderſtāding, as the forme of the
key to the ward of the locke ; for the later, there
is allowed vs an vſe of Reaſon, and argument, ſe-
condarie and reſpectiue ; although not originall
and abſolute : For after the Articles and principles of Religion are placed and exempted from examination of reaſon : It is then permitted vnto vs to
make deriuations and inferences from, and according to the Analogie of them, for our better direction. In Nature this holdeth not, for both the principles are examinable by Induction, though not by a
Medium or *Sillogiſme*: and beſides thoſe principles
or firſt poſitions, haue noe diſcordance with that
reaſon which draweth downe and diduceth the inferiour poſitions. But yet it holdeth not in Religion
alone, but in many knowledges both of greater and
ſmaller Nature, namely wherin there are not onely
Poſita but *Placita*, for in ſuch there can be noe vſe of
abſolute reaſon, we ſee it familiarly in Games of wit,
as Cheſſe, or the like ; The Draughts and firſt lawes
of the Game are poſitiue, but how ? meerely *ad
placitum*, and not examinable by reaſon ; But then
how to direct our play thereupon with beſt aduantage to winne the game, is artificiall and rationall.
So in Humane lawes, there be many groundes and
Maximes, which are *Placita Iuris*, *Poſitiue* vpon authoritie and not vpon reaſon, and therefore not to
be diſputed : But what is moſt iuſt, not abſolutely,
but relatiuely, and according to thoſe Maximes,
that affordeth a long field of diſputation. Such therfore

fore is that fecōdarie reafon, which hath place in diuinitie, which is grounded vpon the *Placets* of God.

Here therefore I note this deficience, that there
hath not bin to my vnderftanding fufficiently enquired & handled, *The true limits and vfe of reafon in*
fpirituall things : as a kinde of diuine Dialectique,
which for that it is not done, it feemeth to me a thing
vfuall, by pretext of true conceiuing that, which is
reuealed, to fearch and mine into that which is not
reuealed, and by pretext of enucleating inferences
and contradictories, to examine that which is pofitiue : The one fort falling into the Error of *Nicode-*
mus, demanding to haue things made more fenfible
then it pleafeth God to reueale them ; *Quomo-*
do pofsit homo nafci cum fit fenex? The other fort into
the Error of the Difciples, which were fcandalized
at a fhew of contradiction: *Quid eft hoc quod dicit no-*
bis, modicum, & non videbitis me, & iterum modicum,
& videbitis me &c.

De vfu le-
gittimo ra-
tionis hu-
manæ in
diuinis.

Vpon this I haue infifted the more, in regard of the
great and bleffed vfe thereof, for this point well laboured and defined of, would in my iudgement be
an *Opiate* to ftaie and bridle not onely the vanitie of
curious fpeculatiōs, wherewith the fchooles labour
but the furie of cōtrouerfies, wherwith the church
laboureth. For it cannot but open mens eyes to fee
that many controuerfies doe meerely pertaine
to that which is either not reuealed or pofitiue, and
that many others doe growe vpon weake and obfcure Inferences or deriuations which latter

fort

sort of men would reuiue the blessed stile of that great Doctor of the Gentiles , would bee carryed thus : *Ego, non Dominus* , and againe *Secundum consilium meum,* in Opinions and counsells, and not in positions and oppositions. But Men are nowe ouer readie to vsurpe the stile. *Non Ego , sed Dominus,* and not so only, but to binde it with the thunder and denunciation of *Curses,* and *Anathemaes,* to the terror of those which haue not sufficiently learned out of *Salomon* , that *The causelesse Curse shall not come.*

Diuinitie hath two principall parts : The matter informed or reuealed : and the nature of the Information or Reuelation : and with the later wee will beginne : because it hath most coherence with that which wee haue now last handled. The nature of the information consisteth of three braunches : The limites of the information; the sufficiencie of the information ; and the acquiring or obtaining the information. Vnto the limits of the information belong these considerations : howe farre forth particular persons continue to bee inspired : how farre forth the Church is inspired: and howe farre forth reason may be vsed ; the last point wherof I haue noted as deficient. Vnto the sufficiency of the information belong two considerations , what points of Religion are foundamentall, & what perfectiue, beeing matter of surder building and perfection vpon one, and the same foundation : and againe how the gradations of light according to the

dispensation

difpenfation of times, are materiall to the fufficien-
cie of beleefe.

Here againe I may rather giue it in aduife, then
note it as deficient, that the points foundamentall,
and the points of further perfection onely ought to
bee with piety and wifedome diftinguifhed : a fubiect tending to much like ende, as that I noted before : for as that other were likely to abate the nomber of controuerfies . So this is like to abate the
heate of manie of them.   Wee fee *Mofes* when he
fawe the *Ifraelite* and the *Egyptian* fight, hee did not
fay, *Why ftriue you* ? but drew his fworde, and flewe
the *Egyptian* : But when hee fawe the two *Ifraelites*
fight, hee faid, *You are brethren , why ftriue you* ? If the
point of doctrine bee an *Egyptian*, it muft bee flaine
by the fword of the fpirit, and not reconciled. But
if it be an *Ifraelite*, though in the wrong: then *Why
ftriue you.* We fee of the foundamentall points, our
Sauiour penneth the league thus, *Hee that is not with
vs is againft vs*, but of points not fundamentall, thus
Hee that is not againft vs, is with vs.   So wee fee the
Coate of our Sauiour was entier without feame,
and fo is the Doctrine of the Scriptures in it felfe :
But the garmente of the Churche was of diuers
colours, and yet not deuided : wee fee the chaffe
may and ought to be feuered from the corne in the
Eare : But the Tares may not be pulled vp from the
corne in the field; So as it is a thing of great vfe well
to define, what, and of what latitude thofe points
are, which doe make men inceiely aliens and dif-

incorporate from the Church of God.

For the obtaining of the information, it resteth vpon the true & sound Interpretation of the Scriptures which are the fountaines of the water of life. The Interpretations of the Scriptures are of two sorts: Methodical, and Solute, or at large, for this diuine water which excelleth so much that of *Iacobs* Well, is drawne forth much in the same kinde, as Naturall Water vseth to bee out of Wells and Fountaines: either it is first forced vp into a Cesterne and from thence fetcht and deriued for vse: or else it is drawne and receiued in Buckets and Vessels immediately where it springeth. The former sort whereof though it seeme to bee the more readie, yet in my iudgement is more subiect to corrupt. This is that Methode which hath exhibited, vnto vs the scholasticall diuinitie, whereby diuinity hath bin reduced into an Art, as into a Cesterne, & the streames of doctrine or positions fetcht and deriued from thence.

In this, Men haue sought three things, a summarie breuitie, a compacted strength, and a compleate perfection: whereof the two first they faile to finde, and the last they ought not to seeke. For as to breuitie, wee see in all summarie Methodes, while men purpose to abridge, they giue cause to dilate. For the summe or abridgement by contraction becommeth obscure, the obscuritie requireth exposition and the exposition is diduced into large comentaries, or into common places, and titles, which growe to be more vast then the originall writings

tings, whence the fumme was at firft extracted. So
we fee the volumes of the fchoole men are greater
much then the firft writings of the fathers, whence
the Maifter of the fentéces made his fumme or collection. So in like manner the volumes of the modern Doctors of the Ciuil Law exceed thofe of the
ancient Iurifconfults, of which *Tribonian* compiled
the digeft. So as this courfe of fummes & cómenta-
ries is that which doth infallibly make the body of
Sciéces more immenfe in quantitie, and more bafe
in fubftance.

And for ftrength, it is true, that knowledges reduced into exact Methodes haue a fhew of ftrength, in
that each part feemeth to fupport & fuftaine the o-
ther: But this is more fatiffactorie then fubftantiall,
like vnto buildings, which ftand by Architecture,
and compaction, which are more fubiect to ruine,
then thofe that are built more ftronge in their feuerall parts though leffe compacted. But it is plaine,
that the more you recede from your grounds, the
weaker doe you conclude, & as in nature, the more
you remoue your felfe from particulars, the greater
peril of Error you doe incur: So, much more in Diuinitie, the more you recede from the Scriptures by
inferences and confequences, the more weake and
dilute are your pofitions.

And as for perfection, or compleatnes in diuinitie
it is not to be fought, which makes this courfe of
Artificiall diuinitie the more fufpecte: For hee
that will reduce a knowledge into an Art, will
make it round and uniforme: But in Diuinitie
many

manie things muſt bee left abrupt and concluded with this : *O altitudo Sapientiæ & ſcientiæ Dei, quam incomprehenſibilia ſunt Iudicia eius, & non inueſtigabiles viæ eius ?* So againe the Apoſtle ſaith, *Ex parte ſcimus ,* and to haue the forme of a totall , where there is but matter for a part, cannot bee without ſupplies by ſuppoſition and preſumption. And therefore I conclude , that the true vſe of theſe Summes and Methods hath place in Inſtitutions or Introductions , preparatorie vnto knowledge: but in them, or by diducement from them, to handle the mayne bodie and ſubſtance of a knoweledge ; is in all Sciences preiudiciall, and in Diuinitie dangerous.

As to the Interpretation of the Scriptures ſolute and at large, there haue beene diuers kindes introduced & deuiſed , ſome of them rather curious and vnſafe , then ſober and warranted. Notwithſtáding thus much muſt be confeſſed,that the Scriptures being giuen by inſpiration , and not by humane reaſon , doe differ from all other books in the Author : which by conſequence doth drawe on ſome difference to be vſed by the Expoſitor. For the Inditer of them did knowe foure things which noe man attaines to knowe,which are the miſteries of the kingdome of glorie ; the perfection of the Lawes of Nature : the ſecrets of the heart of Man: and the future ſucceſſion of all ages. For as to the firſt, it is ſaid. *He that preſſeth into the light, ſhall be oppreſſed of the Glorie .* And againe, *Noe man ſhall ſee*

my

my face and liue. To the second, *When he prepared the heauens I was present, when by lawe and compasse he enclosed the deepe.* To the third, *Neither was it needefull that any should beare witnesse to him of Man, for he knewe well what was in Man.* And to the last, *From the beginning are knowne to the Lord all his workes.*

From the former two of these haue beene drawne certaine senses and expositions of Scriptures, which had need be contained within the bounds of sobrietie; The one *Anagogicall,* and the other *Philosophicall.* But as to the former, Man is not to preuent his time; *Videmus nunc per speculum in Aenigmate, tunc autem facie ad faciem,* wherein neuertheleffe there seemeth to be a libertie graunted, as farre forth as the polishing of this glasse, or some moderate explication of this *Aenigma.* But to presse too farre into it cannot but cause a dissolution and ouerthrowe of the spirite of man. For in the body there are three degrees of that we receiue into it: *Aliment Medecine* and *Poyson* whereof *Aliment* is that which the Nature of Man can perfectly alter & ouercom :Medecine is that which is partly conuerted by Nature, & partly conuerteth nature:& Poyson is that which worketh wholy vpon Nature without that, that nature can in any part worke vpon it. So in the minde whatsoeuer knowledge reason cannot at all worke vpon & conuert, is a neere intoxication and i dangereth a dissolution of the mind & vnderstanding.

But for the liuer, it hath beene extreamely set on foote of late time by the Schoole of *Paracelsus,* and

Ggg some

some others, that haue pretended to finde the truth
of all naturall Philosophy in the Scriptures; scan-
dalizing and traducing all other Philosophie. as
Heathenish and Prophane: But there is noe such
enmitie betweene Gods word, and his workes.
Neither doe they giue honour to the Scriptures, as
they suppose, but much imbase them. For to secke
heauen and earth in the word of God, Whereof it
is saide, *Heauen and Earth shall passe, but my worde
shall not passe,* is to secke temporary things amongst
eternall; And as to secke Diuinitie in Philosophy,
is to secke the liuing amongst the dead; So to secke
Philosophy in Diuinitie is to seek the dead amongst
the liuing; Neither are ye *Pots* or *Lauers*, whose place
was in the outward part of the Temple to be sought
in the holiest place of all, where the Arke of the
testimonie was seated. And againe the scope or
purpose of the spirit of God is not to expresse matters of Nature in the Scriptures, otherwise then in
passage, and for application to mans capacitie and
to matters morall or Diuine. And it is a true Rule,
Authoris aliud agentis parua authoritas. For it were a
strange conclusion, if a man should vse a similitude
for ornament or illustration sake, borrowed from
Nature or historie, according to vulgar conceit, as
of a *Basiliske*, an *Vnicorne*, a *Centaure*, a *Briareus*, an
Hydra or the like, that therefore hee must needes
bee thought to affirme the matter thereof positiuely
to be true; To conclude therefore these two Interpretations,

pretations, the one by reduction or Aenigmaticall, the other Philofophicall or Phificall, which haue beene receiued and purſued in imitationof the *Rabbins* and *Cabalifts*, are to be confined with a *Noli altum ſapere,ſed time*.

But the two later points knowne to God, and vnknowne to Man ; *touching the ſecrets of the heart, and the ſucceſsions of time :* doth make a iuſt and ſound difference betweene the manner of the expoſition of the Scriptures : and all other bookes. For it is an excellent obſeruation which hath beene made vpon the anſweres of our Sauiour Chriſt to many of the queſtions which were propounded to him , how that they are impertinent to the ſtate of the queſtion demanded,the reaſon whereof is, becauſe not being like man , which knowes mans thoughts by his words,but knowing mans thoughts immediately, hee neuer anſwered their words,but their thoughts : much in the like manner it is with the Scriptures,which being written to the thoughts of men,and to the ſucceſsion of all ages,with a foreſight of all hereſies,coatradictions, differing eſtates of the Church,yea,and particularly of the elect, are not to beinterpreted only according to the latitude of the properſenſe of the place , and reſpectiuely towardes that preſent occaſion , whereupon the wordes were vttered ; or in preciſe congruitie or contexture with the wordes before or after , or in contemplation of the principall ſcope of the place, G gg 2 but

but haue in themſelues not onely totally, or colle-
ctiuely, but diſtributiuely in clauſes and wordes,
infinite ſprings and ſtreames of doctrine to water the Church in euerie part, and therefore as the
literall ſenſe is as it were the maine ſtreame or Riuer: So the Morall ſenſe chiefely, and ſometimes
the *Allegoricall* or *Typicall* are they whereof the
Church hath moſt vſe: not that I wiſh men to be
bold in *Allegories*, or *Indulgent* or light in Alluſions:
but that I doe much condemne that Interpretation
of the Scripture, which is onely after the manner as
Men vſe to interprete a prophane booke.

   In this part touching the expoſition of the Scriptures, I can report noe deficience; but by way of remembrance this I will adde, In peruſing Bookes
of Diuinitie, I finde many Bookes of controuerſies,
and many of common places and treatiſes, a maſſe
of poſitiue Diuinitie, as it is made an Arte: a number of Sermons and Lectures, and many prolixe
commentaries vpon the Scriptures with harmonies
and concordances: but that forme of writing in Diuinitie, which in my Iudgement is of all others moſt
rich and precious; is poſitiue Diuinitie collected
vpon particular Texts of Scriptures in briefe obſeruations, not dilated into common places: not chaſe-
ing after controuerſies, not reduced into Methode
of Art, a thing abounding in Sermons, which will
vaniſh, but defectiue in Bookes which wil remaine,
and a thing wherin this age excelleth. For I am per-
ſwaded, and I may ſpeake it, with an *Abſit inuidia*
verbo

verbo, and no waies in derogation of Antiquitie but as in a good emulation betweene the vine and the oliue, That if the choise, and best of those obser-uations vpon Texts of Scriptures which haue beene made dispersedly in Sermons within this your Maiesties Ilands of *Brittanie* by the space of these fortie yeares and more (leauing out the largenesse of exhortations and applications thereupon) had beene set downe in a continuance, it had beene the bestworke in Diuinitie, which had beene written since the Apostles times.

Emanatio-nes Scrip-turarum, in doctrinas Positiuas.

The matter informed by Diuinitie is of two kinds, matter of beliefe, and truth of opinion : and matter of seruice, and adoration ; which is also iudged and directed by the former : The one being as the internall soule of Religion, & the other as the exter-nall body thereof: & therfore the heathen Religion was not onely a worship of Idolls, but the whole Religion was an Idoll in it selfe, for it had noe soule that is, no certaintie of belief or confeßion, as a man may well thinke, considering the chiefe Doctors of their Church were the Poets, and the reason was, becaufe the heathen Gods were noe Iealous Gods, but were glad to be admitted into part, as they had reaso. Neither did they respect the purenesse of harr, so they mought haue externall honour and rites.

But out of these two doe result and issue foure maine branches of Diuinitie ; *Faith, Manners, Ly-turgie*, and *Gouernment* : *Faith* containeth the Do-ctrine of the Nature of G O D, of the attributes of

God

GOD, and of the workes of GOD; The nature
of GOD confifteth of three perfons in vnitie
of GOD head; The attributes of GOD are either common to the deitie, or refpectiue to the perfons ; The workes of GOD fummarie are two,
that of the *Creation* , and that of the *Redemption*:
And both thefe workes, as in Totall they appertaine
to the vnitie of the God-head: So in their parts they
referre to the three-perfons : That of the Creation
in the Maffe of the Matter to the father, in the difpofition of the forme to the Sonne, and in the continuance and conferuation of the being to the Holy
fpirit: So that of the Redemptiõ, in the election and
counfell to the Father, in the whole Act and confū-
mation, to the Sonne: and in the application to the
Holy fpirit : for by the Holy Ghoft was Chrift conceiued in flefh, and by the Holy Ghoft are the Elect
regenerate in fpirite. This worke likewife we confider either effectually in the Elect, or priuately in the
reprobate, or according to apparance in the vifible
Church.

For manners, the Doctrine thereof is contained
in the lawe, which difclofeth finne. The lawe it felfe
is deuided according to the edition thereof, into the
lawe of Nature, the lawe Morall, and the lawe Pofitiue; and according to the ftile, into Negatiue and
Affirmatiue, Prohibitions and Commandements.
Sinne in the matter and fubiect thereof is deuided
according to the Commandements , in the forme
thereof it referreth to the three perfons in deitie,

Sinnes

Sinnes of Infirmitie againſt the father, whoſe more ſpeciall attribute is Power : Sinnes of Ignorance againſt the Sonne, whoſe attribute is wiſedome: and ſinnes of Malice againſt the Holy Ghoſt, whoſe attribute is Grace or Loue . In the motions of it, it either mooueth to the right hand or to the left, either to blinde deuotion, or to prophane &libertine tranſgreſſiõ, either in impoſing reſtraint, where G O D granteth libertie, or in taking libertie where G O D impoſeth reſtrainte . In the degrees and progreſſe of it, it deuideth it ſelfe into thought, word, or Act. And in this part I commend much the diducing of the Lawe of G O D to caſes of conſcience, for that I take indeede to bee a breaking, and not exhibiting whole of the bread of life . But that which quickneth both theſe Doctrines of faith and Manners is the eleuation and conſent of the heart, whereunto appertaine bookes of exhortation, holy meditation, chriſtian reſolution, and the like.

For the Lyturgie or ſeruice, it conſiſteth of the reciprocall Acts betweene G O D and Man, which on the part of G O D are the Preaching of the word and the Sacraments, which are ſeales to the couenant, or as the viſible worde : and on the part of Mans Inuocation of the name of G O D, and vnder the law: Sacrifices, which were as viſible praiers or confeſſions, but now the adoration being in *ſpiritu & veritate* there remaineth only *vituli labiorum,* although

although the vſe of holy vowes of thankefulneſſe and retribution, may be accounted alſo as ſealed petitions.

And for the Gouernment of the Church, it conſiſteth of the patrimonie of the church, the franchiſes of the Church, and the offices, and iuriſdiᵏtions of the Church, and the Lawes of the Church directing the whole : All which haue two conſiderations; the one in them ſelues : the other how they ſtand compatible and agreeable to the Ciuill Eſtate.

This matter of Diuinitie is handled either in forme of inſtruction of truth : or in forme of confutation of falſhood. The declinations from Religion beſides the primitiue which is Atheiſme and the Branches thereof, are three, *Hereſies, Idolatrie,* and *Witch-craft, Hereſies,* when we ſerue the true G O D with a falſe worſhip. *Idolatrie,* when wee worſhip falſe Gods, ſuppoſing them to be true: and *Witch-craft,* when wee adore falſe Gods knowing them to be wicked and falſe. For ſo your Maieſtie doth excellently well obſerue, that *Witchcraft* is the height of *Idolatry.* And yet we ſee thogh theſe be true degrees, *Samuel* teacheth us that they are all of a nature, when there is once a receding from the word of G O D, for ſo he ſaith, *Quaſi Peccatum ariolandi eſt repugnare, & quaſi ſcelus Idololatriæ nolle acquieſcere.*

Theſe thinges I haue paſſed ouer ſo briefely becauſe I can report noe defieience concerning them.

them : For I can finde no fpace or ground that lieth vacant and vnfowne in the matter of Diuinitie, fo diligent haue men beene,either in fowing of
good fee de,or in fowing of Tares.

Thus haue I made as it were a fmall Globe of
the Intellectuall world, as truly and faithfully as
I coulde difcouer, with a note and defcription of
thofe parts which feeme to mee, not conftantly
occupate, or not well conuerted by the labour of
Man. In which , if I haue in any point receded
from that which is commonly receiued, it hath
beene with a purpofe of proceeding in *melius,* and
not in *aliud* : a minde of amendment and proficience; and not of change and difference. For I
could not bee true and conftant to the argument I
handle,if I were not willing to goe beyond others,
but yet not more willing , then to haue others goe
beyond mee againe: which may the better appeare
by this that I haue propounded my opinions naked and vnarmed,not feeking to preoccupate the libertie of mens iudgements by confutations. For in
any thing which is well fet downe , I am in good
hope,that if the firft reading mooue an obiection,
the fecond reading will make an anfwere. And in
thofe things wherein I haue erred, I am fure I haue
not preiudiced the right by litigious arguments;
which certainly haue this contrarie effect and operation, that they adde authoritie to error, and deftroy the authoritie of that which is well inuented.

H h h   For

Of the aduancement of learning.

For queſtion is an honour and preferment to falſhood, as on the other ſide it is a repulſe to truth.
But the errors I claime and challenge to my ſelfe
as mine owne. The good, if any bee, is due
Tanquam adeps ſacrificij, to be incenſed to the honour firſt of the diuine Maieſtie, and next
of your Maieſtie, to whom on
earth I am moſt
bounden.

IN the second Booke, though the words of Art (in Capitall letters) haue seuerall magnitudes, or sizes : therin is meant no difference. The faults in printing, are as follow.

Fol.	Pag.	Lin.		read	
A 2	1	b	17 motions	read	notions
A 3	2	a	29 profection		perfection
E	16	b	5 morall		moderne
E 3	18	a	13 labor that then		labor then
		b	10 diefy		Deifie
			15 *Oue*		*One*
F	20	a	30 *sodilitatem*		*soliditatem*
F 2	21	b	12 as		or (as
F 3	22	b	30 Counsells		Consulls
F 4	23	b	25 *antiquita*		*antiquitas*
G	24	a	8 *Pappia*		*Papia*
I 2	33	a	28 Sciences		Princes
K	36	a	4 great		rare
L 4	43	b	8 face		force
			28 exceed the sences		exceed the pleasures (of the sences
M	44	b	3 to this		to this tend
M 2	45	b	1 with		with any
			6 things		things must

THE SECOND BOOKE.

				read	
D d	14	a	28 portion		partition
E e	17	a	9 Complemet		complement
I i	35	a	24 feuerely		feueredly
K k	39	b	9 iudged		are iudged
M m	48	a	2 moratlitie		moralitie
N n	52	b	6 in these		*in Thesi*
P p	57	b	11 *Sophisme*		*Syllogisme*
P p 4	60	a	22 another tongue		mother tongue
Q q	62	a	2 *Inuentions*		*Inuention*
S s	69	a	5 *Elucendus*		*eliciendus*
	71	b	24 labors?		labors.
T t	72	b	4 or in		or
	73	a	7 sensualitie		sensualitie.

T t 2	72 a	3 Anathemized	Anathematized
	72 b	4 and	an
Vu	74 a	26 *the Æqualitie*	*Æqualitie*
		27 *Euils*	*Euill*
	77 a	3 partie	part
X x	79 b	2 propertie	properly
		19 mattes	matters
	77 b	1 in politikes	politikes
Yy 1	74 b	6 feeme	ferue
		9 vifit	infift
Yy 2	74 a	9 different	diffident
		12 on	in
		20 eafily	eafie
	69 a	30 it	as it
Zz 1	87 b	26 affection	affectation
	79 a	6 hour	fourme
		7 name	it
	89 a	15 *Cicero*	*Q. Cicero*
	b	5 concluded	commended
A aa	91 b	4 *gerit*	*generat*
	92 a	29 many	may
	b	2 gaine	action
		10 is	becaufe it is
	93 a	25 *inutile*	*miffile*
B b b	99 b	24 *rinacur*	*rimatur*
C c c		21 *Calumniari*	*calumniare*
	94 a	21 and : By	and by
E e e	106 a	2 fame	fands
		26 facility	futilitie
	b	18 *infpied*	*infpired*
	b	18 part	port
Fff	111 b	1 of	it
G g g	117 b	15 primitiue	priuatiue

And thefe be marginall notes omitted.

Vu 77 a 15 *De cautelis & malis artibus.*

Zz 79 a 29 *De negotijs gerendis.*

K k 37 a 9 *Pars Phyfiognomiæ, de geftu fiue motu corporis.*

In some few Bookes, in F f: fol. 21. and the beginning of the second page thereof, there is somewhat misplaced, and to be read thus.

an Axiome aswell of iustice, as of the Mathematikes ? And, is there not a true coincidence betweene Commutatiue and Distributiue iustice, and Arithmeticall and Geometricall proportion ? Is not that other, &c.